Editorial Policy

§ 1. Lecture Notes aim to report new developments - quickly, informally, and at a high level. The texts should be reasonably self-contained and rounded off. Thus they may, and often will, present not only results of the author but also related work by other people. Furthermore, the manuscripts should provide sufficient motivation, examples and applications. This clearly distinguishes Lecture Notes manuscripts from journal articles which normally are very concise. Articles intended for a journal but too long to be accepted by most journals, usually do not have this "lecture notes" character. For similar reasons it is unusual for Ph. D. theses to be accepted for the Lecture Notes series.

§ 2. Manuscripts or plans for Lecture Notes volumes should be submitted (preferably in duplicate) either to one of the series editors or to Springer- Verlag, Heidelberg . These proposals are then refereed. A final decision concerning publication can only be made on the basis of the complete manuscript, but a preliminary decision can often be based on partial information: a fairly detailed outline describing the planned contents of each chapter, and an indication of the estimated length, a bibliography, and one or two sample chapters - or a first draft of the manuscript. The editors will try to make the preliminary decision as definite as they can on the basis of the available information.

§ 3. Final manuscripts should preferably be in English. They should contain at least 100 pages of scientific text and should include
- a table of contents;
- an informative introduction, perhaps with some historical remarks: it should be accessible to a reader not particularly familiar with the topic treated;
- a subject index: as a rule this is genuinely helpful for the reader.

Further remarks and relevant addresses at the back of this book.

Lecture Notes in Mathematics   1635

Editors:
A. Dold, Heidelberg
F. Takens, Groningen

Springer
Berlin
Heidelberg
New York
Barcelona
Budapest
Hong Kong
London
Milan
Paris
Santa Clara
Singapore
Tokyo

Emmanuel Hebey

Sobolev Spaces on Riemannian Manifolds

 Springer

Author

Emmanuel Hebey
Département de Mathématiques
Université de Cergy-Pontoise
2 Avenue Adolphe Chauvin
F-95302 Cergy-Pontoise Cedex, France
e-mail: hebey@u-cergy.fr

Cataloging-in-Publication Data applied for

Die Deutsche Bibliothek - CIP-Einheitsaufnahme

Hebey, Emmanuel:
Sobolev spaces on Riemannian manifolds / Emmanuel Hebey. -
Berlin ; Heidelberg ; New York ; Barcelona ; Budapest ; Hong
Kong ; London ; Milan ; Paris ; Santa Clara ; Singapore ;
Tokyo : Springer, 1996
  (Lecture notes in mathematics ; 1635)
  ISBN 3-540-61722-1
NE: GT

Mathematics Subject Classification (1991): 46E35, 58G03

ISSN 0075-8434
ISBN 3-540-61722-1 Springer-Verlag Berlin Heidelberg New York

Typesetting: Camera-ready T_EX output by the author
SPIN: 10479853   46/3142-54321   - Printed on acid-free paper

A Giovanna et Isabelle

Table of contents

Introduction

This monograph is devoted to the study of Sobolev spaces in the general setting of Riemannian manifolds. In addition to being very interesting mathematical structures in their own right, Sobolev spaces play a central role in many branches of mathematics. While analysis proves more and more to be a very powerful means for solving geometrical problems, it is striking that no global study of these spaces exists in the general context of Riemannian manifolds. The objective of this monograph is to fill this gap, at least partially. In so doing, it is intended to serve as a textbook and reference for graduate students and researchers. This monograph also hopes to convince the reader that the naive idea that what is valid for Euclidean spaces must be valid for Riemannian manifolds is completely false. Indeed, as one will see, several surprising phenomena appear when studying Sobolev spaces in the Riemannian context. Elementary questions now give rise to sophisticated developments, where the geometry of the manifolds plays a central role. This monograph is full of such examples.

In a certain sense, Sobolev spaces are studied here for their own interest. Needless to say, they are fundamental in the study of PDE's. A striking example where they have played a major role in the Riemannian context is given by the famous Yamabe problem. The concept of best constants appeared there as crucial for solving limiting cases of some partial differential equations. (Geometric problems often lead to limiting cases of known problems in analysis). While the theory of Sobolev spaces for (non compact) manifolds has its origin in the 70's with the works of Aubin and Cantor, many of the results presented in this monograph have been obtained in the '80's and '90's. As the reader will easily be convinced, the study of Sobolev spaces in the Riemannian context is a field currently undergoing great development !

This monograph presupposes a preliminary course in Riemannian geometry. Not much is assumed to be known so that chapter 1 of Aubin [Au6] should provide specialists in analysis who do not know Riemannian geometry with sufficient knowledge for what follows. Needless to say, many excellent books on Riemannian geometry exist. Although the following ones are not the only possible quality choices, we refer the reader to Chavel [Ch], Gallot-Hulin-Lafontaine [GaHL], Jost [Jo], Kobayashi-Nomizu [KoN], and Spivak [Sp] for more details on what is assumed to be known here.

The material is organized into five chapters and several new results are presented. More precisely, the plan of this monograph is as follows.

Chapter 1 is devoted to the presentation of recent developments of Anderson and Anderson-Cheeger concerning harmonic coordinates, as well as the presentation of a packing result that will be often used in the following chapters.

Chapter 2 is devoted to the presentation of Sobolev spaces on Riemannian manifolds, and to the study of density problems.

Chapter 3 is devoted to Sobolev embeddings. This includes the presentation of general results on the topic, and the study of Sobolev embeddings for Euclidean spaces, compact manifolds, and complete manifolds.

Chapter 4 is devoted to what is currently called the best constants problems. Several results are discussed here, including those concerning the resolution of Aubin's conjecture by Hebey-Vaugon.

Finally, chapter 5 is devoted to the study of the influence of symmetries on Sobolev embeddings.

It is my pleasure and privilege to express my deep thanks to my friend Michel Vaugon for his valuable comments and suggestions about the manuscript. It is also my pleasure and privilege to express my deep thanks to Ms Thanh-Ha Le Thi, and more generally to the staff of Springer-Verlag, for its patience and dedication.

Emmanuel Hebey

June 1996

Chapter 1

Geometric preliminaries

This chapter is devoted to the presentation of geometric results we will use in the sequel. This includes a brief introduction to Riemannian geometry, the presentation of recent results of Anderson and Anderson-Cheeger concerning the harmonic radius of a Riemannian manifold, and the presentation of a packing lemma that will be often used in the following chapters.

1.1 A BRIEF INTRODUCTION TO RIEMANNIAN GEOMETRY

A Riemannian manifold (M, g) is a manifold M together with a $(2, 0)$ tensor field g such that for any x in M, $g(x)$ is a scalar product on $T_x(M)$. Let $|.|_g$ be the norm on $T_x(M)$ with respect to $g(x)$. One can define a distance d_g on M and a positive Radon measure $f \rightarrow \int_M f dv(g)$. Basically, $d_g(x, y)$ is the infimum of the lengths $L(\gamma)$ of all piecewise C^1 curves $\gamma : [a, b] \rightarrow M$ from x to y, where

$$L(\gamma) = \int_a^b |(\frac{d\gamma}{dt})_t|_g dt ,$$

while the Riemannian volume element is given in any chart by

$$dv(g) = \sqrt{det(g_{ij})} dx ,$$

where the g_{ij}'s are the components of g in the chart, and dx is the Lebesgue's volume element of $\mathbf{R}^n$, $n = dim M$. One can also define the Levi-Civita connection of g as the unique linear connection on M which is torsion free and which is such that the covariant derivative of g is zero. The Christoffel symbols of the Levi-Civita connection are then given in any chart by

$$\Gamma_{ij}^k = \frac{1}{2}(\partial_i g_{mj} + \partial_j g_{mi} - \partial_m g_{ij}) g^{mk}$$

where (g^{ij}) denotes the inverse matrix of (g_{ij}), and the Einstein's summation convention is adopted. The components in any chart of the Riemann curvature $Rm_{(M,g)}$ of (M, g), viewed as a $(4, 0)$ tensor field, are given by the relation

$$R_{ijkl} = g_{i\alpha}(\partial_k \Gamma_{jl}^\alpha - \partial_l \Gamma_{jk}^\alpha + \Gamma_{k\beta}^\alpha \Gamma_{jl}^\beta - \Gamma_{l\beta}^\alpha \Gamma_{jk}^\beta)$$

Similarly, the components in any chart of the Ricci curvature $Rc_{(M,g)}$ of (M, g), $Rc_{(M,g)}$ is a $(2, 0)$ tensor field, are given by the relation $R_{ij} = g^{\alpha\beta} R_{i\alpha j\beta}$.

It is well known that sign assumptions on the curvatures give topological
and diffeomorphic informations on the manifold. Striking examples of this fact
are given by Myers theorem (a complete Riemannian manifold whose Ricci curvature satisfies $Rc_{(M,g)} \geq kg$ as bilinear forms, $k > 0$ real, is compact), by the
Cartan-Hadamard theorem (a complete simply connected Riemannian manifold
whose sectional curvature is nonpositive is diffeomorphic to the euclidean space
of same dimension), by the sphere theorem of Berger, Klingenberg, and Rauch
(a compact simply connected Riemannian manifold whose sectional curvature is
1/4 pinched is homeomorphic to the sphere of same dimension), or by Hamil-
ton's theorem (a compact simply connected 3-dimensional Riemannian manifold
whose Ricci curvature is positive is diffeomorphic to the sphere of dimension 3).
On the other hand, by a recent result of Lokhamp, any compact manifold carries
a metric with negative Ricci curvature. Concerning these interactions between
the curvature and the topology of the manifold, one can also think to the Euler caracteristic, which according to the work of Allendoerfer, Chern, Fenchel
and Weil, can be expressed as the integral of some universal polynomial in the
curvature.

Another important object one will meet in the sequel is the injectivity radius. If D denotes the Levi-Civita connection of g, a smooth curve γ is said to
be a geodesic if $D_{(\frac{d\gamma}{dt})}(\frac{d\gamma}{dt}) = 0$. In local coordinates, this means that for any k,

$$(\gamma^k)''(t) + \Gamma_{ij}^k \left(\gamma(t)\right) (\gamma^i)'(t)(\gamma^j)'(t) = 0$$

By the Hopf-Rinow's theorem, any geodesic on a complete Riemannian manifold
(M, g) (that is with respect to the distance d_g) is defined on the whole of $\mathbf{R}$.
Given (M, g) a complete Riemannian manifold and x a point of M, the injectivity
radius $inj_{(M,g)}(x)$ at x is defined as the largest $r > 0$ for which any geodesic γ
of length less than r and having x as an endpoint is minimizing. One has that
$inj_{(M,g)}(x)$ is positive for any x. The injectivity radius of (M, g) is then defined
as the infimum of $inj_{(M,g)}(x)$, $x \in M$. It may be zero.

Closely related to geodesics is the exponential map. Given (M, g) a complete Riemannian n-manifold, x a point of M, and $X \in T_x(M)$, one easily checks
that there exists a unique geodesic γ such that $\gamma(0) = x$ and $(\frac{d\gamma}{dt})_0 = X$. Let
$t \longrightarrow \gamma_X(t)$ be this geodesic. The exponential map exp_x at x is then the map
from $T_x(M)$ to M defined by $exp_x(X) = \gamma_X(1)$. Up to the identification of
$T_x(M)$ with $\mathbf{R}^n$, it is smooth and it defines geodesic normal coordinates at x
on $B_x\left(inj_{(M,g)}(x)\right) = \{y \in M \text{ s.t. } d_g(x,y) < inj_{(M,g)}(x)\}$. More generally, one
can define the cut locus $Cut(x)$ of x, it is a subset of M, and prove that $Cut(x)$
has measure zero, that $inj_{(M,g)}(x) = d_g\left(x, Cut(x)\right)$, and that exp_x is a diffeomorphism from some star-shaped domain of $T_x(M)$ at 0 onto $M \backslash Cut(x)$. In

particular, this implies that the distance function r to a given point is differentiable almost everywhere, with the additional property that $|\nabla r| = 1$ a.e., where $|\nabla r|$ denotes the norm with respect to g of the first covariant derivative of r.

For more details we refer on one hand to [Au6, chapter 1], where a short interesting introduction to Riemannian geometry can be found, on the other hand to the excellent books of Chavel [Ch], Gallot-Hulin-Lafontaine [GaHL], Jost [Jo], Kobayashi-Nomizu [KoN], and Spivak [Sp]. Needless to say, these are not the only possible quality choices.

Once and for all,

ALL THE MANIFOLDS IN THIS MONOGRAPH ARE ASSUMED TO BE
CONNECTED, SMOOTH, WITHOUT BOUNDARY, AND OF DIMENSION $n \geq 3$.

In the following, the Einstein's summation convention is adopted so that $\alpha_i x^i = \sum_i \alpha_i x^i$.

1.2 ESTIMATES ON THE COMPONENTS OF THE METRIC TENSOR

The purpose of this paragraph is to recall how one gets bounds on the components of the metric tensor from bounds on the curvature and the injectivity radius. In other words, how one can choose suitable coordinates so that the components g_{ij} of the metric in these coordinates are bounded in terms of bounds on the curvature and the injectivity radius. The first results in this direction were obtained by using geodesic normal coordinates. In such coordinates, one easily obtains C^0-bounds on the g_{ij}'s from lower and upper bounds on the sectional curvature, and from a lower bound on the injectivity radius. (See [Au6, chapter 1]. See also lemma 1.4 below for further developments). Independently, we know by now that harmonic coordinates are more adapted to this goal. These coordinates were first used by Einstein, then by Lanczos who observed that they simplify the formula for the Ricci tensor. Namely, in such coordinates,

$$R_{ij} = -\frac{1}{2} g^{mk} \partial_{km} g_{ij} \; + \; \text{terms involving at most one derivative of the metric}$$

where (g^{ij}) is the inverse matrix of (g_{ij}) and the R_{ij}'s denote the components of the Ricci curvature of the Riemannian manifold (M, g). Although we are not going to discuss that here, we mention that such a formula has many interesting consequences. We refer to DeTurck-Kazdan [DK] for some of them.

From now on, let Δ_g be the Laplace operator associated to g acting on functions. In local coordinates,

$$\Delta_g u = -g^{ij} \left(\partial_{ij} u - \Gamma_{ij}^k \partial_k u \right) = -\frac{1}{\sqrt{det(g_{ij})}} \partial_m \left(\sqrt{det(g_{ij})} g^{mk} \partial_k u \right)$$

where the Γ_{ij}^k's are the Christoffel symbols of the Levi-Civita connection, and where $det(g_{ij})$ stands for the determinant of the matrix (g_{ij}). One then has the following definition of harmonic coordinates.

Definition 1.1: *A coordinate chart $(x^1, \ldots, x^n)$ on a Riemannian n-manifold (M, g) is called harmonic if $\Delta_g x^k = 0$ for all $k = 1, \ldots, n$. Since $\Delta_g x^k = g^{ij}\Gamma_{ij}^k$, we get that a coordinate chart $(x^1, \ldots, x^n)$ is harmonic if and only if for any $k = 1, \ldots, n$, $g^{ij}\Gamma_{ij}^k = 0$.*

It is easy to prove that for any $x \in M$, there is a neighborhood of x in which harmonic coordinates exist. The proof of such a claim is based on the classical fact that there always exists a smooth solution of $\Delta_g u = 0$ with $u(x)$ and $\partial_i u(x)$ prescribed. The solutions y^j, $j = 1, \ldots, n$, of

$$\begin{cases} \Delta_g y^j = 0 \\ y^j(x) = 0 \\ \partial_i y^j(x) = \delta_i^j \end{cases}$$

are then the desired harmonic coordinates. Furthermore, since composing with linear transformations do not affect the fact that coordinates are harmonic, one easily sees that we can choose the harmonic coordinate system such that $g_{ij}(x) = \delta_{ij}$ for any $i, j = 1, \ldots, n$. We refer to DeTurck-Kazdan [DK] for more basic material on harmonic coordinates.

Let us now define the concept of harmonic radius.

Definition 1.2: *Let (M, g) be a Riemannian n-manifold and let $x \in M$. Given $Q > 1$, $k \in \mathbb{N}$, and $\alpha \in (0, 1)$, we define the $C^{k,\alpha}$ harmonic radius at x as the largest number $r_H = r_H(Q, k, \alpha)(x)$ such that on the geodesic ball $B_x(r_H)$ of center x and radius r_H, there is a harmonic coordinate chart such that the metric tensor is $C^{k,\alpha}$ controlled in these coordinates. Namely, if g_{ij}, $i, j = 1, \ldots, n$, are the components of g in these coordinates, then*

1) $Q^{-1}\delta_{ij} \leq g_{ij} \leq Q\delta_{ij}$ as bilinear forms

2) $\displaystyle\sum_{1 \leq |\beta| \leq k} r_H^{|\beta|} sup_y |\partial_\beta g_{ij}(y)| + \sum_{|\beta|=k} r_H^{k+\alpha} sup_{y \neq z} \frac{|\partial_\beta g_{ij}(z) - \partial_\beta g_{ij}(y)|}{d_g(y,z)^\alpha} \leq Q-1$

where d_g is the distance associated to g. The harmonic radius $r_H(Q, k, \alpha)(M)$ of (M, g) is now defined by $r_H(Q, k, \alpha)(M) = inf_{x \in M} r_H(Q, k, \alpha)(x)$.

One easily checks that the function $x \rightarrow r_H(Q,k,\alpha)(x)$ is 1-lipschitzian on M, since by definition, for any $x, y \in M$,

$$r_H(Q,k,\alpha)(y) \geq r_H(Q,k,\alpha)(x) - d_g(x,y)$$

According to what we have said above, one then gets that the harmonic radius is positive for any fixed smooth compact Riemannian manifold. Now, theorem 1.3 below shows that one obtains lower bounds on the harmonic radius in terms of bounds on the Ricci curvature and the injectivity radius. Roughly speaking, when changing from geodesic normal coordinates to harmonic coordinates, one controls the components of the metric in terms of the Ricci curvature instead of the whole Riemann curvature. As it is stated below, theorem 1.3 can be found in Hebey-Herzlich [HH], and we refer to [HH] for its proof. For original references, we refer to Anderson [An2], and Anderson-Cheeger [AC]. (See also Jost-Karcher [JK]). Let (M,g) be a Riemannian manifold. In the following, $Rc_{(M,g)}$ denotes its Ricci curvature, $\nabla^j Rc_{(M,g)}$ denotes the jth-covariant derivative of $Rc_{(M,g)}$, and if x is some point of M, $inj_{(M,g)}(x)$ denotes the injectivity radius of (M,g) at x.

Theorem 1.3: *Let $\alpha \in (0,1)$, $Q > 1$, $\delta > 0$. Let (M,g) be a Riemannian n-manifold, and Ω an open subset of M. Set*

$$\Omega(\delta) = \{x \in M \text{ s.t. } d_g(x, \Omega) < \delta\}$$

where d_g is the distance associated to g. Suppose that for some $\lambda \in \mathbf{R}$ and $i > 0$, we have that for all $x \in \Omega(\delta)$,

$$Rc_{(M,g)}(x) \geq \lambda g_x \quad \text{and} \quad inj_{(M,g)}(x) \geq i$$

Then there exists a positive constant $C = C(n, Q, \alpha, \delta, i, \lambda)$, depending only on n, Q, α, δ, i, and λ, such that for any $x \in \Omega$, $r_H(Q, 0, \alpha)(x) \geq C$. In addition, if instead of the bound $Rc_{(M,g)}(x) \geq \lambda g_x$ we assume that for some $k \in \mathbf{N}$ and some positive constants $C(j)$,

$$|\nabla^j Rc_{(M,g)}(x)| \leq C(j) \quad \text{for all } j = 0, \ldots, k \text{ and all } x \in \Omega(\delta)$$

then, there exists a positive constant $C = C(n, Q, k, \alpha, \delta, i, C(j)_{0 \leq j \leq k})$, depending only on n, Q, k, α, δ, i, and the $C(j)$'s, $0 \leq j \leq k$, such that for any $x \in \Omega$, $r_H(Q, k+1, \alpha)(x) \geq C$.

The proof of theorem 1.3 is by contradiction. It is too long to be developed here. Let us just say that the general idea is to construct a sequence of Riemannian n-manifolds with harmonic radius less than or equal to 1, to prove that it converges to the euclidean space $\mathbf{R}^n$, and to get the contradiction by noting that this would imply that the harmonic radius of $\mathbf{R}^n$ is less than or equal to 1. (Obviously, $\mathbf{R}^n$ has an infinite harmonic radius).

As already mentioned, analogous estimates to those of theorem 1.3 are available if one works with geodesic normal coordinates instead of harmonic coordinates. These estimates are rougher since, for instance, they involve the Riemann curvature instead of the Ricci curvature. Anyway, such results are sometimes useful, and, actually, lemma 1.4 below is used in the proof of theorem 4.12 of chapter 4. This is why we mention it. For details on its proof, we refer to Hebey-Vaugon [HV3, section III]. Let (M, g) be a Riemannian manifold. In the following, $Rm_{(M,g)}$ denotes its Riemann curvature (viewed as a (4,0) tensor field), $\nabla Rm_{(M,g)}$ denotes the first covariant derivative of $Rm_{(M,g)}$, and, as above, $inj_{(M,g)}(x)$ denotes the injectivity radius of (M, g) at x.

Lemma 1.4: *Let (M, g) be a Riemannian n-manifold. Suppose that for some $x \in M$ there exist positive constants Λ_1 and Λ_2 such that $|Rm_{(M,g)}| \leq \Lambda_1$ and $|\nabla Rm_{(M,g)}| \leq \Lambda_2$ on the geodesic ball $B_x\big(inj_{(M,g)}(x)\big)$ of center x and radius $inj_{(M,g)}(x)$. Then there exist positive constants $K = K(n, \Lambda_1, \Lambda_2)$ and $\delta = \delta(n, \Lambda_1, \Lambda_2)$, depending only on n, Λ_1 and Λ_2, such that the components g_{ij} of g in geodesic normal coordinates at x satisfy: for any $i, j, k = 1, \ldots, n$ and any $y \in B_0^e\Big(min\big(\delta, inj_{(M,g)}(x)\big)\Big)$,*

(i) $\frac{1}{4}\delta_{ij} \leq g_{ij}\big(exp_x(y)\big) \leq 4\delta_{ij}$ (as bilinear forms)

(ii) $|g_{ij}\big(exp_x(y)\big) - \delta_{ij}| \leq K|y|^2$ and $|\partial_k g_{ij}\big(exp_x(y)\big)| \leq K|y|$

where for $t > 0$, $B_0^e(t)$ denotes the euclidean ball of $\mathbf{R}^n$ with center 0 and radius t, and $|y|$ is the euclidean distance from 0 to y. In addition, one has that

$$\lim_{(\Lambda_1, \Lambda_2) \to (0,0)} \delta(n, \Lambda_1, \Lambda_2) = +\infty \quad and \quad \lim_{(\Lambda_1, \Lambda_2) \to (0,0)} K(n, \Lambda_1, \Lambda_2) = 0$$

The proof of lemma 1.4 starts with standard estimates of the theory of Jacobi fields. (See for instance [Au6, chapter 1]). Then it relies on a careful study of the formula for the curvature in polar coordinates.

1.3 FROM LOCAL ANALYSIS TO GLOBAL ANALYSIS

The purpose of this paragraph is to prove a packing lemma that will be used many times in the following chapters. This lemma is by now classical. It was first proved by Calabi (unpublished) under the assumptions that the sectional

curvature of the manifold is bounded and that the injectivity radius of the man-
ifold is positive. (See Aubin [Au2], and Cantor [Can]). By Croke's result [Cr, proposition 14] it was then possible to replace the assumption on the sectional curvature by a lower bound on the Ricci curvature. Finally, by an ingenious use of Gromov's theorem, theorem 1.5 below, one obtains the result under the more general form of lemma 1.6. When we will discuss Sobolev inequalities on complete manifolds, this lemma will be an important tool in the process of passing from local to global inequalities.

The following result, due to Gromov [GrLP], is by now classical. We refer the reader to [GaHL, theorem 4.19] for details on its proof.

Theorem 1.5: *Let (M,g) be a complete Riemannian n-manifold whose Ricci curvature satisfies $Rc_{(M,g)} \geq (n-1)kg$ for some $k \in \mathbf{R}$. Then, for any $0 < r < R$ and any $x \in M$,*

$$Vol_g\big(B_x(R)\big) \leq \frac{V_k(R)}{V_k(r)} Vol_g\big(B_x(r)\big)$$

where $Vol_g\big(B_x(t)\big)$ denotes the volume of the geodesic ball of center x and radius t, and where $V_k(t)$ denotes the volume of a ball of radius t in the complete simply connected Riemannian n-manifold of constant curvature k. In particular, for any $r > 0$ and any $x \in M$, $Vol_g\big(B_x(r)\big) \leq V_k(r)$.

Remark: Let b_n be the volume of the Euclidean ball of radius one. It is well known (see for instance [GaHL]) that for any $t > 0$,

$$V_{-1}(t) = n b_n \int_0^t (sinh s)^{n-1} ds$$

where, according to the notations of theorem 1.5, $V_{-1}(t)$ denotes the volume of a ball of radius t in the simply connected hyperbolic space of dimension n. It is then easy to prove that for any $k \geq 0$ and any $t > 0$,

$$b_n t^n \leq V_{-k}(t) \leq b_n t^n e^{(n-1)\sqrt{k}t}$$

One just has to note that for $s \geq 0$, $s \leq sinh s \leq se^s$, and that if $g' = \alpha^2 g$ are Riemannian metrics on a n-manifold M, where α is some positive real number, then for any $x \in M$ and any $t > 0$,

$$Vol_{g'}\big(B'_x(t)\big) = \alpha^n Vol_g\big(B_x(t/\alpha)\big)$$

As a consequence, by theorem 1.5 and what we just said we get that if (M,g) is a complete Riemannian n-manifold whose Ricci curvature satisfies $Rc_{(M,g)} \geq kg$ for some $k \in \mathbf{R}$, then for any $x \in M$ and any $0 < r < R$,

$$Vol_g\big(B_x(R)\big) \leq e^{\sqrt{(n-1)|k|}R}\Big(\frac{R}{r}\Big)^n Vol_g\big(B_x(r)\big)$$

This explicit inequality will be used sometimes in the sequel.

Let (M, g) be a Riemannian manifold. For any $x \in M$ and any $r > 0$, we denote by $B_x(r)$ the geodesic ball of center x and radius r. Independently, we say that a family (Ω_k) of open subsets of M is a uniformly locally finite covering of M if the following holds: $\cup_k \Omega_k = M$ and there exists an integer N such that each point $x \in M$ has a neighborhood which intersects at most N of the Ω_k's.

Lemma 1.6: *Let (M, g) be a complete Riemannian n-manifold with Ricci curvature bounded from below by some $k \in \mathbf{R}$, and let $\rho > 0$ be given. There exists a sequence (x_i) of points of M such that for any $r \geq \rho$:*

(i) the family $\big(B_{x_i}(r)\big)$ is a uniformly locally finite covering of M, and there is an upper bound for N in terms of n, ρ, r, and k

(ii) for any $i \neq j$, $B_{x_i}(\rho/2) \cap B_{x_j}(\rho/2) = \emptyset$

Proof of lemma 1.6: By theorem 1.5 and the remark following this theorem we get that for any $x \in M$ and any $0 < r < R$,

$$Vol_g\big(B_x(r)\big) \geq e^{-\sqrt{(n-1)|k|}R}\Big(\frac{r}{R}\Big)^n Vol_g\big(B_x(R)\big) \qquad (1)$$

Independently, we claim that there exists a sequence (x_i) of points of M such that

$$M = \bigcup_i B_{x_i}(\rho) \quad \text{and} \quad \forall i \neq j, \ B_{x_i}(\rho/2) \cap B_{x_j}(\rho/2) = \emptyset \qquad (2)$$

Actually, let

$$X_\rho = \big\{(x_i)_I, \ x_i \in M, \text{ s.t. } I \text{ is countable and } \forall i \neq j, \ d_g(x_i, x_j) \geq \rho\big\}$$

where d_g is the Riemannian distance associated to g. X_ρ is partially ordered by inclusion and, obviously, every chain in X_ρ has an upper bound. Hence, by Zorn's lemma, X_ρ contains a maximal element (x_i), and (x_i) satisfies (2).

Let (x_i) be such that (2) is satisfied. For $r > 0$ and $x \in M$ we define

$$I_r(x) = \big\{i \text{ s.t. } x \in B_{x_i}(r)\big\}$$

By (1) we get that for $r \geq \rho$

$$Vol_g\big(B_x(r)\big) \geq \frac{1}{2^n} e^{-2\sqrt{(n-1)|k|}r} Vol_g\big(B_x(2r)\big)$$

$$\geq \frac{1}{2^n} e^{-2\sqrt{(n-1)|k|}r} \sum_{i \in I_r(x)} Vol_g\big(B_{x_i}(\rho/2)\big)$$

since

$$\bigcup_{i \in I_r(x)} B_{x_i}(\rho/2) \subset B_x(2r) \quad \text{and} \quad B_{x_i}(\rho/2) \cap B_{x_j}(\rho/2) = \emptyset \text{ if } i \neq j$$

But, again by (1),

$$Vol_g\big(B_{x_i}(\rho/2)\big) \geq e^{-2\sqrt{(n-1)|k|}r}\Big(\frac{\rho}{4r}\Big)^n Vol_g\big(B_{x_i}(2r)\big)$$

and since for any $i \in I_r(x)$, $B_x(r) \subset B_{x_i}(2r)$, we get that

$$Vol_g\big(B_x(r)\big) \geq \Big(\frac{\rho}{8r}\Big)^n e^{-4\sqrt{(n-1)|k|}r} Card I_r(x) Vol_g\big(B_x(r)\big)$$

where $Card$ stands for the cardinality. As a consequence, for any $r \geq \rho$ there exists $C = C(n, \rho, r, k)$ such that for any $x \in M$, $Card I_r(x) \leq C$.

Now, let $B_{x_i}(r)$ be given, $r \geq \rho$, and suppose that N balls $B_{x_j}(r)$ have a nonempty intersection with $B_{x_i}(r)$, $j \neq i$. Then, obviously, $Card I_{2r}(x_i) \geq N+1$. Hence,

$$N \leq C(n, \rho, 2r, k) - 1$$

and this proves the lemma.

Chapter 2

Sobolev spaces

This short chapter is devoted to the presentation of Sobolev spaces on Riemannian manifolds and to the study of density problems. For Sobolev spaces on open subsets of $\mathbf{R}^n$, we refer the reader to the very complete books of Adams [Ad], and Maz'ja [Maz].

2.1 FIRST DEFINITIONS

Let (M, g) be a Riemannian manifold. For k an integer and $u \in C^\infty(M)$, $\nabla^k u$ denotes the kth covariant derivative of u (with the convention $\nabla^0 u = u$). As an example, the components of ∇u in local coordinates are given by $(\nabla u)_i = \partial_i u$, while the components of $\nabla^2 u$ in local coordinates are given by

$$(\nabla^2 u)_{ij} = \partial_{ij} u - \Gamma_{ij}^k \partial_k u$$

By definition one has that

$$|\nabla^k u|^2 = g^{i_1 j_1} \dots g^{i_k j_k} (\nabla^k u)_{i_1 \dots i_k} (\nabla^k u)_{j_1 \dots j_k}$$

For k an integer and $p \geq 1$ real, we denote by $C_k^p(M)$ the space of smooth functions $u \in C^\infty(M)$ such that $|\nabla^j u| \in L^p(M)$ for any $j = 0, \dots, k$. Hence,

$$C_k^p(M) = \left\{ u \in C^\infty(M) \text{ s.t. } \forall j = 0, \dots, k, \int_M |\nabla^j u|^p dv(g) < \infty \right\}$$

where, in local coordinates, $dv(g) = \sqrt{\det(g_{ij})} dx$, and where dx stands for the Lebesgue's volume element of $\mathbf{R}^n$, $n = \dim M$. If M is compact, one has that $C_k^p(M) = C^\infty(M)$ for all k and $p \geq 1$.

Definition 2.1: *The Sobolev space $H_k^p(M)$ is the completion of $C_k^p(M)$ with respect to the norm*

$$\|u\|_{H_k^p} = \sum_{j=0}^{k} \left(\int_M |\nabla^j u|^p dv(g) \right)^{1/p}$$

Noting that a Cauchy sequence in $C_k^p(M)$ is also a Cauchy sequence in $L^p(M)$, and that a Cauchy sequence in $C_k^p(M)$ wich converges to 0 in $L^p(M)$ converges to 0 in $C_k^p(M)$, the Sobolev spaces $H_k^p(M)$ can be seen as subspaces of

$L^p(M)$. This is the point of vue we adopt in the sequel. More precisely, one can look at $H_k^p(M)$ as the space of functions u in $L^p(M)$ which are limit in $L^p(M)$ of a Cauchy sequence (u_m) in $C_k^p(M)$, and define the norm $\|u\|_{H_k^p}$ as above where $|\nabla^j u|$, $0 \leq j \leq k$, is now the limit in $L^p(M)$ of $|\nabla^j u_m|$. One checks without any difficulty that these spaces are Banach spaces. The following results are then easy to prove.

Proposition 2.2: *If $p = 2$, $H_k^2(M)$ is a Hilbert space when equipped with the equivalent norm*

$$\|u\| = \sqrt{\sum_{j=0}^{k} \int_M |\nabla^j u|^2 dv(g)}$$

The scalar product $\langle ., . \rangle$ associated to $\|.\|$ is defined by

$$\langle u, v \rangle = \sum_{m=0}^{k} \int_M \left(g^{i_1 j_1} \ldots g^{i_m j_m} (\nabla^m u)_{i_1 \ldots i_m} (\nabla^m v)_{j_1 \ldots j_m} \right) dv(g)$$

Proposition 2.3: *If M is compact, $H_k^p(M)$ does not depend on the Riemannian metric.*

Proposition 2.4: *If $p > 1$, $H_k^p(M)$ is reflexive.*

Of course, proposition 2.3 is not anymore true for non-compact manifolds. Think for instance to some non-compact manifold M endowed with two metrics, the volume of M for one of these two metrics being finite, the volume of M for the other one being infinite. Then the constant function $u = 1$ belongs to the Sobolev spaces associated to the metric of finite volume, while it does not belong to the Sobolev spaces associated to the metric of infinite volume. Independently, let us now mention that as an easy consequence of Meyers-Serrin's theorem [Ad, theorem 3.16], one has the following. (When we refer to lipschitzian functions we refer to global lipschitzian functions).

Lemma 2.5: *Let (M, g) be a Riemannian manifold and $u : M \to \mathbf{R}$ a lipschitzian function on M which equals zero outside a compact subset of M. Then $u \in H_1^p(M)$ for any $p \geq 1$.*

Proof of lemma 2.5: Let $u : M \to \mathbf{R}$ be a lipschitzian function on M which equals zero outside a compact subset K of M. Let $(\Omega_k, \phi_k)_{k=1,\ldots,N}$ be a finite number of charts such that $K \subset \cup_{k=1}^{N} \Omega_k$ and such that for any $k = 1, \ldots, N$,

$$\phi_k(\Omega_k) = B_0^c(1) \quad \text{and} \quad \frac{1}{R}\delta_{ij} \leq g_{ij}^k \leq R\delta_{ij} \text{ as bilinear forms}$$

where $R > 1$ is some real number, $B_0^e(1)$ is the unit euclidean ball of center 0 in $\mathbf{R}^n$, and the g_{ij}^k's are the components of g in (Ω_k, ϕ_k). Let $(\alpha_k)_{k=1,\dots,N}$ be a smooth partition of unity subordinate to the covering $(\Omega_k)_{k=1,\dots,N}$. One easily gets that for any $k = 1, \dots, N$, $(\alpha_k u) \circ \phi_k^{-1}$ is a lipschitzian function on $B_0^e(1)$ with respect to the euclidean distance. As a consequence, $(\alpha_k u) \circ \phi_k^{-1}$ is differentiable almost everywhere (by Rademacher's theorem), with the additional property that the partial derivatives of $(\alpha_k u) \circ \phi_k^{-1}$ belong to $L^p(B_0^e(1))$ for any $p \geq 1$ (since they are bounded), and that they define the weak derivatives of $(\alpha_k u) \circ \phi_k^{-1}$ (by the absolute continuity of lipschitzian functions one any line segment). By Meyers-Serrin's theorem one then gets that $\alpha_k u \in H_1^p(M)$ for any $p \geq 1$ and any $k = 1, \dots, N$. Since $u = \sum_{k=1}^N \alpha_k u$, this ends the proof of the lemma.

More generally, recall here that for Ω a domain in $\mathbf{R}^n$, a function $u : \Omega \to \mathbf{R}^n$ is weakly differentiable in Ω if and only if it is equal almost everywhere to a function $\bar{u}$ that is absolutely continuous on almost all line segments in Ω parallel to the coordinate axes and whose partial derivatives (which consequently exist a.e.), are locally integrable in Ω. If in addition $\bar{u}$ and its partial derivatives belong to $L^p(\Omega)$, then by Myers-Serrin's theorem $u \in H_1^p(\Omega)$.

2.2 DENSITY PROBLEMS

From now on, let $\mathcal{D}(M)$ be the space of C^∞ functions with compact support in M. We define the Sobolev space $\overset{o}{H}_k^p(M)$ as follows.

Definition 2.6: *The Sobolev space $\overset{o}{H}_k^p(M)$ is the closure of $\mathcal{D}(M)$ in $H_k^p(M)$.*

For the euclidean space $\mathbf{R}^n$, it is well known that $\overset{o}{H}_k^p(\mathbf{R}^n) = H_k^p(\mathbf{R}^n)$ (see for instance [Ad, theorem 2.4]). A natural and interesting problem is then to determine for which manifolds one has $\overset{o}{H}_k^p(M) = H_k^p(M)$. Needless to say, the question becomes interesting when M is non-compact. As shown by Aubin [Au2], the density of $\mathcal{D}(M)$ in $H_k^p(M)$ holds for complete manifolds when $k = 1$ (theorem 2.7 below). On the other hand, the situation is more complicate when $k \geq 2$ and we need some assumptions on the manifold. In [Au2], Aubin proved that for any $p \geq 1$ and $k \geq 2$, $\mathcal{D}(M)$ is dense in $H_k^p(M)$ if (M, g) is complete with positive injectivity radius and Riemann curvature bounded up to the order $k - 2$. We prove here that Aubin's result still holds if one replaces the assumptions on the Riemann curvature by analoguous assumptions on the Ricci curvature (theorem 2.8 below). First, we recall how one proves the following.

Theorem 2.7: *If (M,g) is complete, then, for any $p \geq 1$, $\overset{o}{H_1^p}(M) = H_1^p(M)$.*

Proof of theorem 2.7: Let $f : \mathbf{R} \to \mathbf{R}$ be defined by

$$f(t) = 1 \text{ when } t \leq 0, \quad f(t) = 1 - t \text{ when } 0 \leq t \leq 1, \quad f(t) = 0 \text{ when } t \geq 1$$

and let $u \in C_1^p(M)$ where $p \geq 1$ is some given real number. Let x be some point of M and set

$$u_j(y) = u(y)f\big(d_g(x,y) - j\big)$$

where d_g is the Riemannian distance associated to g, $j \in \mathbf{N}$, and $y \in M$. By lemma 2.5, $u_j \in H_1^p(M)$ for any j, and since $u_j = 0$ outside a compact subset of M, one easily gets that for any j, $u_j \in \overset{o}{H_1^p}(M)$. (Just note that if $(u_m) \in C_1^p(M)$ converges to u_j in $H_1^p(M)$, and if $\alpha \in \mathcal{D}(M)$, then (αu_m) converges to αu_j in $H_1^p(M)$. Then, choose $\alpha \in \mathcal{D}(M)$ such that $\alpha = 1$ where $u_j \neq 0$). Independently, one clearly has that for any j,

$$\left(\int_M |u_j - u|^p dv(g)\right)^{1/p} \leq \left(\int_{M \backslash B_x(j)} |u|^p dv(g)\right)^{1/p}$$

and

$$\left(\int_M |\nabla(u_j - u)|^p dv(g)\right)^{1/p}$$
$$\leq \left(\int_{M \backslash B_x(j)} |\nabla u|^p dv(g)\right)^{1/p} + \left(\int_{M \backslash B_x(j)} |u|^p dv(g)\right)^{1/p}$$

where $B_x(j)$ is the geodesic ball of center x and radius j. Hence, $\lim_{j \to \infty} u_j = u$ in $H_1^p(M)$. This ends the proof of the theorem.

Theorem 2.7 is false for Ω a bounded open subset of $\mathbf{R}^n$ endowed with the euclidean metric e. One easily checks that $\overset{o}{H_1^2}(\Omega) \neq H_1^2(\Omega)$. Consider for this purpose the scalar product $\langle .,. \rangle$ of proposition 2.2 (with $g = e$ and $k = 1$), and let $u \in C^\infty(\Omega) \cap H_1^2(\Omega)$ be such that $\Delta_e u + u = 0$, $u \not\equiv 0$. (For instance, $u = sinhx_1$, x_1 the first coordinate of x). Then for any $v \in \mathcal{D}(\Omega)$, $\langle u, v \rangle = \int_\Omega (\Delta_e u + u)v dx = 0$, so that $u \notin \overset{o}{H_1^2}(\Omega)$. In particular, this implies that the constant function 1 does not belong to $\overset{o}{H_1^2}(\Omega)$. Let us now prove what we announced few lines above.

Theorem 2.8: *Let (M,g) be a complete Riemannian manifold with positive injectivity radius and let $k \geq 2$ be an integer. Suppose that there exists a positive* *constant C such that for any $j = 0, \ldots, k - 2$, $|\nabla^j Rc_{(M,g)}| \leq C$. Then for any $p \geq 1$, $\overset{\mathrm{o}}{H}{}^p_k(M) = H^p_k(M)$.*

Proof of theorem 2.8: Suppose that $inj_{(M,g)} > 0$ and that there exists $C > 0$ such that for any $j = 0, \ldots, k - 2$, $|\nabla^j Rc_{(M,g)}| \leq C$. By theorem 1.3 one has that for any real numbers $Q > 1$ and $\alpha \in (0,1)$, the harmonic radius $r_H = r_H(Q, k - 1, \alpha)$ is positive. Fix for instance $Q = 4$ and $\alpha = 1/2$. (α will play no role in the following of the proof). For any $x \in M$ one then has that there exists a harmonic chart $\phi : B_x(r_H) \to \mathbf{R}^n$ such that the points 1) and 2) of definition 1.2 are satisfied with $Q = 4$ and $\alpha = 1/2$. (Without loss of generality we can also assume that $\phi(x) = 0$). In particular, we get that for any $r \leq r_H$

$$B^e_0(r/2) \subset \phi\big(B_x(r)\big) \subset B^e_0(2r)$$

where $B^e_0(r/2)$ (resp. $B^e_0(2r)$) denotes the euclidean ball of center 0 and radius $r/2$ (resp. $2r$). Let $\beta \in \mathcal{D}(\mathbf{R}^n)$ be such that

$$0 \leq \beta \leq 1 \,, \quad \beta = 1 \text{ on } B^e_0(r_H/8) \,, \quad \beta = 0 \text{ on } \mathbf{R}^n \backslash B^e_0(r_H/4)$$

As a consequence of the inclusions above we get that $\beta \circ \phi \in \mathcal{D}(M)$ satisfies

$$0 \leq \beta \circ \phi \leq 1 \,, \quad \beta \circ \phi = 1 \text{ on } B_x(r_H/16) \,, \quad \text{and } \beta \circ \phi = 0 \text{ on } M \backslash B_x(r_H/2)$$

From now on let (x_i) be a sequence of points of M such that $M = \cup_i B_{x_i}(r_H/16)$ and such that the covering $\big(B_{x_i}(r_H/2)\big)$ is uniformly locally finite. The existence of such a sequence is given by lemma 1.6. Let $\phi_i : B_{x_i}(r_H) \to \mathbf{R}^n$ be as above and set $\beta_i = \beta \circ \phi_i$. Since the components of the metric tensor are C^{k-1}-controlled in the charts $(B_{x_i}(r_H), \phi_i)$, one easily gets that there exists $C > 0$ such that for any i and any $m = 0, \ldots, k$, $|\nabla^m \beta_i| \leq C$. Let us now set $\eta_i = \beta_i / \sum_j \beta_j$. As a consequence of what we have said above, (η_i) is a smooth partition of unity subordinate to the covering $\big(B_{x_i}(r_H/2)\big)$, and since this covering is uniformly locally finite, one easily obtains that there exists some constant $\tilde{C} \geq 1$ such that for any $m = 0, \ldots, k$, $\sum_i |\nabla^m \eta_i| \leq \tilde{C}$. Now, fix $u \in C^p_k(M)$ where $p \geq 1$ is some given real number. Proposition 2.8 will obviously be proved if we show that for any $\epsilon > 0$ there exists $u_0 \in \mathcal{D}(M)$ such that $\|u - u_0\|_{H^p_k} < \epsilon$. Fix $\epsilon > 0$ and let $\Omega \subset M$ be some bounded subset of M such that

$$\sum_{m=0}^{k} C^{m+1}_{k+1} \left(\int_{M \backslash \Omega} |\nabla^m u|^p \, dv(g) \right)^{1/p} < \epsilon/\tilde{C}$$

where $\tilde{C}$ is as above and

$$C_{k+1}^{m+1} = \frac{(k+1)!}{(m+1)!(k-m)!}$$

Since the covering $\left(B_{x_i}(r_H/2)\right)$ is uniformly locally finite, one easily obtains that there exists some integer N such that for any $i \geq N+1$, $B_{x_i}(r_H/2) \cap \Omega = \emptyset$. Set $u_0 = (1-\eta)u$ where $1 - \eta = \sum_{i=1}^{N} \eta_i$. Then $u_0 \in \mathcal{D}(M)$ and

$$\|u - u_0\|_{H_k^p} \leq \sum_{m=0}^{k} \|\nabla^m(\eta u)\|_p$$

where $\|f\|_p = \left(\int_M |f|^p dv(g)\right)^{1/p}$. But

$$|\nabla^m(\eta u)| \leq \sum_{j=0}^{m} C_m^j |\nabla^j \eta| |\nabla^{m-j} u|$$

and since $\mathrm{Supp}\,\eta \subset M \backslash \Omega$ and $\sum_i |\nabla^j \eta_i| \leq \tilde{C}$ for any $j = 0, \ldots, k$, we get that

$$\|\nabla^m(\eta u)\|_p \leq \tilde{C} \sum_{j=0}^{m} C_m^j \left(\int_{M \backslash \Omega} |\nabla^j u|^p dv(g)\right)^{1/p}$$

As a consequence, noting that for any $0 \leq m \leq k$, $\sum_{j=m}^{k} C_j^m = C_{k+1}^{m+1}$, we get that

$$\begin{aligned}
\|u - u_0\|_{H_k^p} &\leq \tilde{C} \sum_{m=0}^{k} \sum_{j=0}^{m} C_m^j \left(\int_{M \backslash \Omega} |\nabla^j u|^p dv(g)\right)^{1/p} \\
&= \tilde{C} \sum_{m=0}^{k} \left(\sum_{j=m}^{k} C_j^m\right) \left(\int_{M \backslash \Omega} |\nabla^m u|^p dv(g)\right)^{1/p} \\
&= \tilde{C} \sum_{m=0}^{k} C_{k+1}^{m+1} \left(\int_{M \backslash \Omega} |\nabla^m u|^p dv(g)\right)^{1/p}
\end{aligned}$$

Since

$$\sum_{m=0}^{k} C_{k+1}^{m+1} \left(\int_{M \backslash \Omega} |\nabla^m u|^p dv(g)\right)^{1/p} < \epsilon/\tilde{C}$$

we have shown that for any $\epsilon > 0$ and any $u \in \mathcal{C}_k^p(M)$ there exists $u_0 \in \mathcal{D}(M)$ such that $\|u - u_0\|_{H_k^p} < \epsilon$. As already mentioned, this ends the proof of the theorem.

As a straightforward consequence of theorem 2.8 one has the following.

Corollary 2.9: *For any Riemannian covering $(\tilde{M}, \tilde{g})$ of a compact Riemannian manifold (M, g), for any integer k, and for any $p \geq 1$ real, $\overset{\circ}{H}{}^p_k(\tilde{M}) = H^p_k(\tilde{M})$.*

Finally, we mention that when $k = p = 2$, theorem 2.8 can be improved. More precisely, one has the following.

Proposition 2.10: *Let (M, g) be a complete Riemannian manifold whose Ricci curvature is bounded from below and whose injectivity radius is positive. Then $\overset{\circ}{H}{}^2_2(M) = H^2_2(M)$.*

Proof of proposition 2.10: Let $K^2_2(M)$ be the completion of

$$\tilde{C}^2_2(M) = \{u \in C^\infty(M) \text{ s.t. } u, |\nabla u|, \Delta_g u \in L^2(M)\}$$

with respect to

$$\|u\|_{K^2_2} = \left(\int_M u^2 dv(g)\right)^{1/2} + \left(\int_M |\nabla u|^2 dv(g)\right)^{1/2} + \left(\int_M |\Delta_g u|^2 dv(g)\right)^{1/2}$$

and let $\overset{\circ}{K}{}^2_2(M)$ be the closure of $\mathcal{D}(M)$ in $K^2_2(M)$. We assume that $Rc_{(M,g)} \geq \lambda g$ for some $\lambda \in \mathbf{R}$, and that $inj_{(M,g)} > 0$. By theorem 1.3 one then has that for any real numbers $Q > 1$ and $\alpha \in (0, 1)$, the harmonic radius $r_H = r_H(Q, 0, \alpha)$ is positive. Noting that in a harmonic coordinate chart, $\Delta_g u = -g^{ij} \partial_{ij} u$ for any $u \in C^\infty(M)$, similar arguments to those used in the proof of proposition 2.8 prove that $\overset{\circ}{K}{}^2_2(M) = K^2_2(M)$. Independently, one clearly has that for any $u \in C^\infty(M)$, $|\Delta_g u|^2 \leq n|\nabla^2 u|^2$. Hence, $H^2_2(M) \subset K^2_2(M)$ and the embedding is continuous. Finally, by the Bochner-Lichnerowicz-Weitzenböck formula [Lic], one has that for any $u \in \mathcal{D}(M)$,

$$\int_M |\nabla^2 u|^2 dv(g) = \int_M |\Delta_g u|^2 dv(g) - \int_M Rc_{(M,g)}(\nabla u, \nabla u) dv(g)$$

$$\leq \int_M |\Delta_g u|^2 dv(g) + |\lambda| \int_M |\nabla u|^2 dv(g)$$

Hence, $\|u\|_{H^2_2} \leq \left(1 + \sqrt{|\lambda|}\right)\|u\|_{K^2_2}$ for any $u \in \mathcal{D}(M)$, and according to what we have just said we get that $\overset{\circ}{H}{}^2_2(M) = \overset{\circ}{K}{}^2_2(M)$. As a consequence,

$$\overset{\circ}{H}{}^2_2(M) \subset H^2_2(M) \subset K^2_2(M) = \overset{\circ}{K}{}^2_2(M) = \overset{\circ}{H}{}^2_2(M)$$

and this ends the proof of the proposition.

Chapter 3

Sobolev embeddings

Let (M, g) be a Riemannian n-manifold. We will be concerned in this chapter with the following question: under what conditions the Sobolev embeddings are valid on M ? Once and for all, when we refer to Sobolev embeddings, we refer to the following.

Sobolev embeddings: For p, q two real numbers with $1 \leq q < p$, and k, m two integers with $0 \leq m < k$, if $1/p = 1/q - (k-m)/n$ then $H_k^q(M) \subset H_m^p(M)$. Here and in the sequel, the notation $H_k^q(M) \subset H_m^p(M)$ includes the continuity of the embedding, namely the existence of a positive constant $C = C(M, p, q, k, m)$ such that for any $u \in H_k^q(M)$, $\|u\|_{H_m^p} \leq C \|u\|_{H_k^q}$.

Such embeddings were first proved by Sobolev [So] for $\mathbf{R}^n$, and the result is now referred to as the Sobolev embedding theorem. We will see later in this chapter that these embeddings are valid for compact manifolds, while the situation is more intricate for complete manifolds.

3.1 PRELIMINARY RESULTS

We prove in this paragraph two basic results that will be often used in the sequel. The first one is well known. It is stated as follows.

Lemma 3.1: *Let (M, g) be a complete Riemannian n-manifold. Suppose that the embedding $H_1^1(M) \subset L^{n/(n-1)}(M)$ is valid. Then, for any real numbers $1 \leq q < p$ and any integers $0 \leq m < k$ satisfying $1/p = 1/q - (k-m)/n$, $H_k^q(M) \subset H_m^p(M)$.*

Proof of lemma 3.1: We prove that if $H_1^1(M) \subset L^{n/(n-1)}(M)$ then, for any $1 \leq q < n$ and $1/p = 1/q - 1/n$, $H_1^q(M) \subset L^p(M)$. We refer to [Au6, proposition 2.11] for the proof that the other embeddings $H_k^q(M) \subset H_m^p(M)$ are also valid. Let $A \in \mathbf{R}$ be such that for any $u \in H_1^1(M)$

$$\left(\int_M |u|^{n/(n-1)} dv(g) \right)^{(n-1)/n} \leq A \int_M \left(|\nabla u| + |u| \right) dv(g)$$

Let $1 \leq q < n$, $1/p = 1/q - 1/n$, and $u \in \mathcal{D}(M)$. Set $\phi = |u|^{p(n-1)/n}$. Applying Hölder's inequality, we get that

$$\left(\int_M |u|^p dv(g) \right)^{(n-1)/n} = \left(\int_M |\phi|^{n/(n-1)} dv(g) \right)^{(n-1)/n}$$

$$\leq A \int_M \left(|\nabla \phi| + |\phi| \right) dv(g)$$

$$= \frac{Ap(n-1)}{n} \int_M |u|^{p'} |\nabla u| dv(g) + A \int_M |u|^{p(n-1)/n} dv(g)$$

$$\leq \frac{Ap(n-1)}{n} \left(\int_M |u|^{p'q'} dv(g) \right)^{1/q'} \left(\int_M |\nabla u|^q dv(g) \right)^{1/q}$$

$$+ A \left(\int_M |u|^{p'q'} dv(g) \right)^{1/q'} \left(\int_M |u|^q dv(g) \right)^{1/q}$$

where $1/q + 1/q' = 1$ and $p' = p(n-1)/n - 1$. But $p'q' = p$ since $1/p = 1/q - 1/n$. As a consequence, for any $u \in \mathcal{D}(M)$,

$$\left(\int_M |u|^p dv(g) \right)^{1/p} \leq \frac{Ap(n-1)}{n} \left\{ \left(\int_M |\nabla u|^q dv(g) \right)^{1/q} + \left(\int_M |u|^q dv(g) \right)^{1/q} \right\}$$

By theorem 2.7, this ends the proof of the lemma.

Remarks: 1) Note that the proof of lemma 3.1 shows that if $A \in \mathbf{R}$ is such that for any $u \in H_1^1(M)$

$$\left(\int_M |u|^{n/(n-1)} dv(g) \right)^{(n-1)/n} \leq A \int_M \left(|\nabla u| + |u| \right) dv(g)$$

then, for any $1 \leq q < n$ and any $u \in H_1^q(M)$,

$$\left(\int_M |u|^p dv(g) \right)^{1/p} \leq \frac{Ap(n-1)}{n} \left\{ \left(\int_M |\nabla u|^q dv(g) \right)^{1/q} + \left(\int_M |u|^q dv(g) \right)^{1/q} \right\}$$

where $1/p = 1/q - 1/n$.

2) We will see in paragraph 3.5 that there exists a kind of converse to lemma 3.1 for complete manifolds with Ricci curvature bounded from below. Namely, we will see that if (M, g) is complete with Ricci curvature bounded from below, and if one has that $H_1^q(M) \subset L^p(M)$ for some $1 < q < n$ and $1/p = 1/q - 1/n$, then $H_1^1(M) \subset L^{n/(n-1)}(M)$.

Let us now discuss the following lemma. This lemma due to Carron [Car2] extends to the whole scale of the embeddings of H_1^q in L^p the well known fact that the validity of the embedding of H_1^1 in $L^{n/(n-1)}$ implies a uniform lower bound for the volume of balls with respect to their center.

Lemma 3.2: *Let (M, g) be a complete Riemannian n-manifold. Suppose that the embedding $H_1^q(M) \subset L^p(M)$ is valid for some $1 \leq q < n$ and $1/p = 1/q - 1/n$.*

Then for any $r > 0$ there exists a positive constant $v = v(M, q, r)$ such that for any $x \in M$, $Vol_g(B_x(r)) \geq v$.

Proof of lemma 3.2: By hypothesis $H_1^q(M) \subset L^p(M)$ for some $1 \leq q < n$ and p satisfying $1/p = 1/q - 1/n$. Let $A > 0$ be such that for any $u \in H_1^q(M)$,

$$\left(\int_M |u|^p dv(g) \right)^{1/p} \leq A \left\{ \left(\int_M |\nabla u|^q dv(g) \right)^{1/q} + \left(\int_M |u|^q dv(g) \right)^{1/q} \right\}$$

Let $r > 0$, let x be some point of M, and let $v \in H_1^q(M)$ be such that $v = 0$ on $M \backslash B_x(r)$. We have by Hölder,

$$\left(\int_M |v|^q dv(g) \right)^{1/q} \leq Vol_g(B_x(r))^{1/n} \left(\int_M |v|^p dv(g) \right)^{1/p}$$

Hence,

$$\frac{1}{Vol_g(B_x(r))^{1/n}} - A \leq A \frac{\left(\int_M |\nabla v|^q dv(g) \right)^{1/q}}{\left(\int_M |v|^q dv(g) \right)^{1/q}}$$

Fix $x \in M$ and let $R > 0$ be given. Then, either $Vol_g(B_x(R)) > (1/2A)^n$, either $Vol_g(B_x(R)) \leq (1/2A)^n$ in which case we get that for any $r \in (0, R]$,

$$\frac{1}{Vol_g(B_x(r))^{1/n}} - A \geq \frac{1}{2Vol_g(B_x(r))^{1/n}}$$

Suppose that $Vol_g(B_x(R)) \leq (1/2A)^n$. We then have that for any $r \in (0, R]$ and any $v \in H_1^q(M)$ such that $v = 0$ on $M \backslash B_x(r)$,

$$\frac{1}{(2A)^q} Vol_g(B_x(r))^{-q/n} \leq \frac{\int_M |\nabla v|^q dv(g)}{\int_M |v|^q dv(g)}$$

From now on, let

$$v(y) = r - d_g(x, y) \text{ if } d_g(x, y) \leq r$$
$$v(y) = 0 \text{ if } d_g(x, y) \geq r.$$

Clearly v is lipschitzian and $v = 0$ on $M \backslash B_x(r)$. Hence, see lemma 2.5, v belongs to $H_1^q(M)$. As a consequence,

$$\frac{1}{(2A)^q} Vol_g(B_x(r))^{-q/n} \leq \frac{Vol_g(B_x(r))}{\int_{B_x(r/2)} v^q dv(g)} \leq \frac{2^q Vol_g(B_x(r))}{r^q Vol_g(B_x(r/2))}$$

and we get that for any $r \leq R$,

$$Vol_g(B_x(r)) \geq \left(\frac{r}{4A} \right)^{nq/(n+q)} Vol_g(B_x(r/2))^{n/(n+q)}$$

By induction we then get that for any $m \in \mathbb{N}\backslash\{0\}$,

$$Vol_g\big(B_x(R)\big) \geq \big(\frac{R}{2A}\big)^{q\alpha(m)}\big(\frac{1}{2}\big)^{q\beta(m)}Vol_g\big(B_x(R/2^m)\big)^{\gamma(m)} \qquad (3)$$

where

$$\alpha(m) = \sum_{i=1}^{m}\big(\frac{n}{n+q}\big)^i \ , \ \beta(m) = \sum_{i=1}^{m}i\big(\frac{n}{n+q}\big)^i \ , \ \text{and } \gamma(m) = \big(\frac{n}{n+q}\big)^m$$

But, see for instance [GaHL, theorem 3.98], $Vol_g\big(B_x(r)\big) = b_n r^n (1+o(r))$ where b_n is the volume of the euclidean ball of radius one. Hence,

$$\lim_{m\to\infty} Vol_g\big(B_x(R/2^m)\big)^{\gamma(m)} = 1$$

In addition, we have that

$$\sum_{i=1}^{\infty}\big(\frac{n}{n+q}\big)^i = \frac{n}{q} \ \text{ and } \ \sum_{i=1}^{\infty}i\big(\frac{n}{n+q}\big)^i = \frac{n(n+q)}{q^2}$$

As a consequence, letting $m \to \infty$ in (3) we get that

$$Vol_g\big(B_x(R)\big) \geq \big(\frac{1}{2^{(n+2q)/q}A}\big)^n R^n$$

Finally, for any $x \in M$ and any $R > 0$,

$$Vol_g\big(B_x(R)\big) \geq \min\big(1/2A, R/2^{(n+2q)/q}A\big)^n$$

and this ends the proof of the lemma.

Remarks: 1) The proof of lemma 3.2 gives the exact dependence of v. Namely, v depends on n, q, r, and the constant A of the embedding of $H_1^q(M)$ in $L^p(M)$.

2) We used the fact that $Vol_g\big(B_x(r)\big) = b_n r^n (1+o(r))$. More precisely, one has

$$Vol_g\big(B_x(r)\big) = b_n r^n \big(1 - \frac{Scal_{(M,g)}(x)}{6(n+2)}r^2 + o(r^2)\big)$$

where $Scal_{(M,g)}$, the trace with respect to g of $Rc_{(M,g)}$, is the scalar curvature of (M,g). See for instance [GaHL, theorem 3.98].

3) More standard arguments, similar to those developed in [Ch, chapter 6], lead to the same result when $q = 1$. More precisely, one can prove that for any $x \in M$ and almost all $r > 0$,

$$Vol_g\big(B_x(r)\big)^{(n-1)/n} \leq A\frac{d}{dr}Vol_g\big(B_x(r)\big) + AVol_g\big(B_x(r)\big)$$

where A is the constant of the embedding of $H_1^1(M)$ in $L^{n/(n-1)}(M)$. As a consequence, for $R > 0$ given, either one has that $Vol_g(B_x(R)) \geq (1/2A)^n$, either one has that $Vol_g(B_x(R)) \leq (1/2A)^n$ and one obtains that for almost all $r \in (0, R]$,

$$(1/2A)Vol_g(B_x(r))^{1-1/n} \leq \frac{d}{dr}Vol_g(B_x(r))$$

Integrating this last inequality one then gets that for any $x \in M$ and any $R > 0$, $Vol_g(B_x(R)) \geq \min((1/2A)^n, (R/2nA)^n)$.

3.2 THE SOBOLEV EMBEDDING THEOREM FOR $\mathbf{R}^n$

The purpose of this paragraph is to recall how one can prove the well known fact that the Sobolev embeddings are valid for $\mathbf{R}^n$. The original proof, given by Sobolev [So], was based on a quite difficult lemma. We present here the proof of Gagliardo[Ga] and Nirenberg [Ni].

Lemma 3.3: *For any $u \in \mathcal{D}(\mathbf{R}^n)$,*

$$\left(\int_{\mathbf{R}^n} |u|^{n/(n-1)}dx\right)^{(n-1)/n} \leq \frac{1}{2}\prod_{i=1}^{n}\left(\int_{\mathbf{R}^n}|\frac{\partial u}{\partial x_i}|dx\right)^{1/n}$$

where dx is the Lebesgue's volume element of $\mathbf{R}^n$.

Proof of lemma 3.3: We present the proof for $n = 3$. The proof for $n \neq 3$ is similar. Let P be a point of $\mathbf{R}^3$, (x, y, z) the coordinates in $\mathbf{R}^3$, (x_0, y_0, z_0) the coordinates of P, and D_x (respectively D_y, D_z) the straight line trough P parallel to the x-axis (respectively y-,z-axis). With these notations, the Lebesgue's volume element dx of the lemma is $dxdydz$. Let $u \in \mathcal{D}(\mathbf{R}^n)$. We then have that

$$u(P) = \int_{-\infty}^{x_0} (\partial_x u)(x, y_0, z_0)dx = -\int_{x_0}^{+\infty} (\partial_x u)(x, y_0, z_0)dx$$

As a consequence, $|u(P)| \leq \frac{1}{2}\int_{D_x} |(\partial_x u)(x, y_0, z_0)|dx$. With similar arguments for $\partial_y u$ and $\partial_z u$ we get that

$$|u(P)|^{3/2} \leq (\frac{1}{2})^{3/2}\left(\int_{D_x} |(\partial_x u)(x, y_0, z_0)|dx\right)^{1/2}$$

$$\times \left(\int_{D_y} |(\partial_y u)(x_0, y, z_0)|dy\right)^{1/2}\left(\int_{D_z} |(\partial_z u)(x_0, y_0, z)|dz\right)^{1/2}$$

Now, integrating x_0 over $\mathbf{R}$ yields, by Hölder's inequality,

$$\int_{D_x} |u(x, y_0, z_0)|^{3/2}dx \leq (\frac{1}{2})^{3/2}\left(\int_{D_x} |(\partial_x u)(x, y_0, z_0)|dx\right)^{1/2}$$

$$\times \left(\int_{D_{xy}} |(\partial_y u)(x, y, z_0)| dx dy \right)^{1/2} \left(\int_{D_{xz}} |(\partial_z u)(x, y_0, z)| dx dz \right)^{1/2}$$

where D_{xy} (resp. D_{xz}) is the plane through P parallel to the x- and y-axis (resp. x- and z-axis). Integrating y_0 over $\mathbf{R}$ then yields, by Hölder's inequality,

$$\int_{D_{xy}} |u(x, y, z_0)|^{3/2} dx dy \le \left(\frac{1}{2} \right)^{3/2} \left(\int_{D_{xy}} |(\partial_x u)(x, y, z_0)| dx dy \right)^{1/2}$$

$$\times \left(\int_{D_{xy}} |(\partial_y u)(x, y, z_0)| dx dy \right)^{1/2} \left(\int_{R^3} |(\partial_z u)(x, y, z)| dx dy dz \right)^{1/2}$$

Finally, integrating z_0 over $\mathbf{R}$, we obtain the inequality of lemma 3.3.

With such a result we are now in position to prove that the Sobolev embeddings are valid for $\mathbf{R}^n$.

Theorem 3.4: *Let $q \in [1, n)$ and let p be such that $1/p = 1/q - 1/n$. For any $u \in H_1^q(\mathbf{R}^n)$,*

$$\left(\int_{R^n} |u|^p dx \right)^{1/p} \le \frac{p(n-1)}{2n} \left(\int_{R^n} |\nabla u|^q dx \right)^{1/q} \tag{4}$$

In particular, for any real numbers $1 \le q < p$ and any integers $0 \le m < k$ satisfying $1/p = 1/q - (k-m)/n$, $H_k^q(\mathbf{R}^n) \subset H_m^p(\mathbf{R}^n)$.

Proof of theorem 3.4: As a straightforward consequence of lemma 3.3, we have that $H_1^1(\mathbf{R}^n) \subset L^{n/(n-1)}(\mathbf{R}^n)$ and that for any $u \in H_1^1(\mathbf{R}^n)$,

$$\left(\int_{R^n} |u|^{n/(n-1)} dx \right)^{(n-1)/n} \le \frac{1}{2} \int_{R^n} |\nabla u| dx$$

By lemma 3.1 one then gets that for any real numbers $1 \le q < p$ and any integers $0 \le m < k$ satisfying $1/p = 1/q - (k-m)/n$, $H_k^q(\mathbf{R}^n) \subset H_m^p(\mathbf{R}^n)$. Finally, a similar computation to the one made in the proof of lemma 3.1 shows that for any $1 \le q < n$ and any $u \in H_1^q(\mathbf{R}^n)$,

$$\left(\int_{R^n} |u|^p dx \right)^{1/p} \le \frac{p(n-1)}{2n} \left(\int_{R^n} |\nabla u|^q dx \right)^{1/q}$$

where $1/p = 1/q - 1/n$.

Remark: The constant $\frac{p(n-1)}{2n}$ in (4) is not optimal. We refer to paragraph 4.2 for the precise value of the best constant K such that for any $u \in H_1^q(\mathbf{R}^n)$,

$$\left(\int_{R^n} |u|^p dx \right)^{1/p} \le K \left(\int_{R^n} |\nabla u|^q dx \right)^{1/q}$$

3.3 SOBOLEV EMBEDDINGS FOR COMPACT MANIFOLDS

First we recall how one can prove the well-known fact that the Sobolev embeddings are valid for compact manifolds. Then we discuss the well-known Rellich-Kondrakov theorem which states that for compact manifolds, and with the exception of extreme cases, all the embeddings given by the Sobolev embedding theorem are compact. Finally, we say some words about the Poincaré and Sobolev-Poincaré inequalities.

Theorem 3.5: *Let (M, g) be a compact Riemannian n-manifold. For any real numbers $1 \leq q < p$ and any integers $0 \leq m < k$ satisfying $1/p = 1/q - (k-m)/n$, $H_k^q(M) \subset H_m^p(M)$.*

Proof of theorem 3.5: By lemma 3.1 we just have to prove that the embedding $H_1^1(M) \subset L^{n/(n-1)}(M)$ is valid. Now, since M is compact, M can be covered by a finite number of charts $(\Omega_m, \phi_m)_{m=1,...,N}$ such that for any m the components g_{ij} of g in (Ω_m, ϕ_m) satisfy $(1/2)\delta_{ij} \leq g_{ij} \leq 2\delta_{ij}$ as bilinear forms. Let (η_m) be a smooth partition of unity subordinate to the covering (Ω_m). For any $u \in C^\infty(M)$ and any m, one then has that

$$\int_M |\eta_m u|^{n/(n-1)} dv(g) \leq 2^{n/2} \int_{R^n} |(\eta_m u) \circ \phi_m^{-1}(x)|^{n/(n-1)} dx$$

and

$$\int_M |\nabla(\eta_m u)| dv(g) \geq 2^{-(n+1)/2} \int_{R^n} |\nabla((\eta_m u) \circ \phi_m^{-1})(x)| dx$$

Independently, by theorem 3.4,

$$\left(\int_{R^n} |(\eta_m u) \circ \phi_m^{-1}(x)|^{n/(n-1)} dx \right)^{(n-1)/n} \leq \frac{1}{2} \int_{R^n} |\nabla((\eta_m u) \circ \phi_m^{-1})(x)| dx$$

for any m. As a consequence, for any $u \in C^\infty(M)$,

$$\left(\int_M |u|^{n/(n-1)} dv(g) \right)^{(n-1)/n} \leq \sum_{m=1}^{N} \left(\int_M |\eta_m u|^{n/(n-1)} dv(g) \right)^{(n-1)/n}$$

$$\leq 2^{n-1} \sum_{m=1}^{N} \int_M |\nabla(\eta_m u)| dv(g)$$

$$\leq 2^{n-1} \int_M |\nabla u| dv(g)$$

$$+ 2^{n-1} \left(\max_M \sum_{m=1}^{N} |\nabla \eta_m| \right) \int_M |u| dv(g)$$

This ends the proof of the theorem.

Let us now discuss the Rellich-Kondrakov theorem. If (M, g) is compact, $Vol_{(M,g)}$ is finite. As a consequence, for any $1 \leq p \leq \tilde{p}$, $L^{\tilde{p}}(M) \subset L^p(M)$. By theorem 3.5 we then get that for any integers $j \geq 0$ and $m \geq 1$, any $q \geq 1$ real, and any p real such that $1 \leq p \leq nq/(n-mq)$, $H^q_{j+m}(M) \subset H^p_j(M)$. The Rellich-Kondrakov theorem then asserts that these embeddings are compact provided $p < nq/(n-mq)$. Recall that if $(E, \|.\|_E)$ and $(F, \|.\|_F)$ are normed spaces with $E \subset F$, the embedding of E in F is said to be compact if bounded subsets of $(E, \|.\|_E)$ are precompact subsets of $(F, \|.\|_F)$. The Rellich-Kondrakov theorem can then be stated as follows.

Theorem 3.6: *Let (M, g) be a compact Riemannian n-manifold. For any integers $j \geq 0$ and $m \geq 1$, any real number $q \geq 1$, and any real number p such that $1 \leq p < nq/(n-mq)$, the embedding of $H^q_{j+m}(M)$ in $H^p_j(M)$ is compact.*

Corollary 3.7: *Let (M, g) be a compact Riemannian n-manifold. For any $1 \leq q < n$ and any $p \geq 1$ such that $1/p > 1/q - 1/n$, the embedding of $H^q_1(M)$ in $L^p(M)$ is compact.*

Proof of theorem 3.6: The proof is very classical so we will just say some words about it. First, see for instance [Au6, theorem 2.33], one can prove that for any bounded domain Ω of $\mathbf{R}^n$, any $1 \leq q < n$, and any $1 \leq p < nq/(n-q)$, the embedding of $\overset{o}{H}{}^q_1(\Omega)$ in $L^p(\Omega)$ is compact. Now, since the composition of two continuous embeddings is compact if one of them is compact, we get that for any bounded domain Ω of $\mathbf{R}^n$ and any m, q, and p as in theorem 3.6, the embedding of $\overset{o}{H}{}^q_m(\Omega)$ in $L^p(\Omega)$ is compact. (Just note that by theorem 3.4, $\overset{o}{H}{}^q_m(\Omega) \subset \overset{o}{H}{}^{q'}_1(\Omega)$ where $1/q' = 1/q - (m-1)/n$, while, according to what we have just said, the embedding of $\overset{o}{H}{}^{q'}_1(\Omega)$ in $L^p(\Omega)$ is compact). Using Meyers-Serrin's theorem ([Ad, theorem 3.16]) it is then easy to prove by finite induction that for any bounded domain Ω of $\mathbf{R}^n$ and any j, m, q, and p as in theorem 3.6, the embedding of $\overset{o}{H}{}^q_{j+m}(\Omega)$ in $\overset{o}{H}{}^p_j(\Omega)$ is compact. Finally, one can proceed as in [Au6, theorem 2.34] to get theorem 3.6.

Let us now say some words about the Poincaré and Sobolev-Poincaré inequalities. First we prove the following.

Lemma 3.8: *Let (M, g) be a compact Riemannian n-manifold, and let $1 \leq q < n$ be a real number. There exists a positive constant $A = A(M, g, q)$ such that for* any $u \in H_1^q(M)$,

$$\left(\int_M |u - \bar{u}|^q dv(g) \right)^{1/q} \leq A \left(\int_M |\nabla u|^q dv(g) \right)^{1/q}$$

where $\bar{u} = \frac{1}{Vol_{(M,g)}} \int_M u \, dv(g)$.

Such inequalities are referred to as Poincaré inequalities.

Proof of lemma 3.8: Suppose first that $q > 1$. To prove lemma 3.8 we just have to prove that

$$\inf_{u \in \mathcal{H}} \int_M |\nabla u|^q dv(g) > 0$$

where

$$\mathcal{H} = \left\{ u \in H_1^q(M) \text{ s.t. } \int_M |u|^q dv(g) = 1 \text{ and } \int_M u \, dv(g) = 0 \right\}$$

Let $(u_k) \in \mathcal{H}$ be such that

$$\lim_{k \to \infty} \int_M |\nabla u_k|^q dv(g) = \inf_{u \in \mathcal{H}} \int_M |\nabla u|^q dv(g)$$

Combining the fact that $H_1^q(M)$ is reflexive for $q > 1$ and the Rellich-Kondrakov theorem, theorem 3.6, there exists a subsequence (u_k) of (u_k) which converges weakly in $H_1^q(M)$ and strongly in $L^q(M) \cap L^1(M)$. Let v be its limit. The strong convergence in $L^q(M) \cap L^1(M)$ implies that $v \in \mathcal{H}$, while we get with the weak convergence that

$$\int_M |\nabla v|^q dv(g) \leq \lim_{k \to \infty} \int_M |\nabla u_k|^q dv(g)$$

As a consequence, $\inf_{u \in \mathcal{H}} \int_M |\nabla u|^q dv(g)$ is attained by v, and since v cannot be constant,

$$\inf_{u \in \mathcal{H}} \int_M |\nabla u|^q dv(g) > 0$$

This proves the Poincaré inequalities for $q > 1$. When $q = 1$ we can use the well-known fact that on a compact manifold there always exists a Green's function for the laplacian. More precisely, see for instance [Au6, theorem 4.13], if (M, g) is a compact Riemannian n-manifold there exists $G : M \times M \to \mathbf{R}$ such that:

(i) for any $u \in C^\infty(M)$ and any $x \in M$,

$$u(x) = \frac{1}{Vol_{(M,g)}} \int_M u \, dv(g) + \int_M G(x, y) \Delta_g u(y) dv_g(y)$$

(ii) $G(x, y) = G(y, x)$ and $G(x, y)$ is C^∞ on $M \times M \backslash \Delta$ where Δ is the diagonal

$$\Delta = \{(x, y) \in M \times M \text{ s.t. } x = y\}$$

(iii) there exists a constant $K > 0$ such that for any $(x, y) \in M \times M \backslash \Delta$,

$$|G(x, y)| \le \frac{K}{r^{n-2}} \text{ and } |\nabla_y G(x, y)| \le \frac{K}{r^{n-1}}$$

where $r = d_g(x, y)$ is the Riemannian distance from x to y.

From now on let $u \in C^\infty(M)$ be such that $\int_M u \, dv(g) = 0$. We then have that for any x,

$$u(x) = \int_M G(x, y) \Delta_g u(y) dv_g(y)$$

Hence,

$$|u(x)| \le \int_M |\nabla_y G(x, y)| |\nabla u(y)| dv_g(y)$$

and

$$\int_M |u(x)| dv_g(x) \le \int_M \int_M |\nabla_y G(x, y)| |\nabla u(y)| dv_g(x) dv_g(y)$$

$$\le C \int_M |\nabla u(y)| dv_g(y)$$

where $C > 0$ is such that for any $y \in M$, $\int_M |\nabla_y G(x, y)| dv_g(x) \le C$. (Recall that G satisfies $|\nabla_y G(x, y)| \le K/r^{n-1}$). As a consequence, for any $u \in C^\infty(M)$ such that $\int_M u \, dv(g) = 0$,

$$\int_M |u| dv(g) \le C \int_M |\nabla u| dv(g)$$

and the Poincaré inequality for $q = 1$ is proved. This ends the proof of the lemma.

Now, if one mixes the Poincaré inequalities with the Sobolev inequalities related to the embeddings $H_1^q \subset L^p$, we get the Sobolev-Poincaré inequalities. Namely, one has the following.

Proposition 3.9: Let (M, g) be a compact Riemannian n-manifold, and let p, q two real numbers such that $1 \le q < n$ and $1/p = 1/q - 1/n$. There exists a positive constant $A = A(M, g, q)$ such that for any $u \in H_1^q(M)$,

$$\left(\int_M |u - \bar{u}|^p dv(g) \right)^{1/p} \le A \left(\int_M |\nabla u|^q dv(g) \right)^{1/q}$$

where $\bar{u} = \frac{1}{Vol_{(M,g)}} \int_M u\, dv(g).$

Proof of proposition 3.9: By theorem 3.5 there exists a positive constant B such that for any $u \in H_1^q(M)$,

$$\left(\int_M |u - \bar{u}|^p \, dv(g) \right)^{1/p} \leq B\left(\int_M |\nabla u|^q dv(g) \right)^{1/q} + B\left(\int_M |u - \bar{u}|^q dv(g) \right)^{1/q}$$

Independently, by lemma 3.8, there exists $C > 0$ such that for any $u \in H_1^q(M)$,

$$\left(\int_M |u - \bar{u}|^q dv(g) \right)^{1/q} \leq C\left(\int_M |\nabla u|^q dv(g) \right)^{1/q}$$

Hence, for any $u \in H_1^q(M)$,

$$\left(\int_M |u - \bar{u}|^p dv(g) \right)^{1/p} \leq B(1 + C)\left(\int_M |\nabla u|^q dv(g) \right)^{1/q}$$

and this proves the proposition.

3.4 SOBOLEV EMBEDDINGS AND CONVERGENCE OF MANIFOLDS

Let (Z, d_Z) be a metric space and let A, B be two subsets of Z. For $\epsilon > 0$ set

$$U_\epsilon(A) = \{z \in Z \text{ s.t. } d_Z(z, A) < \epsilon\} \quad \text{and} \quad U_\epsilon(B) = \{z \in Z \text{ s.t. } d_Z(z, B) < \epsilon\}$$

The classical Hausdorff distance for subsets of a single metric space is then defined by

$$d_Z^H(A, B) = \inf\{\epsilon > 0 \text{ s.t. } B \subset U_\epsilon(A) \text{ and } A \subset U_\epsilon(B)\}$$

More generally, if (X, d_X), (Y, d_Y) are two metric spaces, one can define the Gromov-Hausdorff distance $d_H(X, Y)$ as

$$d_H(X, Y) = \inf d_Z^H\left(f(X), g(Y)\right)$$

where the inf is taken over all metric spaces (Z, d_Z) and all isometries

$$f : (X, d_X) \longrightarrow (Z, d_Z) \ , \ g : (Y, d_Y) \longrightarrow (Z, d_Z)$$

For compact metric spaces, d_H is a distance. Let $n \geq 3$ be an integer, $q \in [1, n)$, $V > 0$, and $A > 0$. We define $\mathcal{M}(n, q, V, A)$ as the set of compact Riemannian n-manifolds (M, g) satisfying $Vol_{(M,g)} \leq V$ and such that for any $u \in C^\infty(M)$,

$$\left(\int_M |u|^{nq/(n-q)} dv(g) \right)^{(n-q)/nq} \leq A\left(\left(\int_M |\nabla u|^q dv(g) \right)^{1/q} + \left(\int_M |u|^q dv(g) \right)^{1/q} \right)$$

One then has the following.

Proposition 3.10: *For any n, $q \in [1,n)$, $V > 0$, and $A > 0$, $\mathcal{M}(n,q,V,A)$ is precompact for the Gromov-Hausdorff distance.*

Proof of proposition 3.10: By lemma 3.2 there exists $v : \mathbf{R}^{+*} \to \mathbf{R}^{+*}$ such that for any $(M,g) \in \mathcal{M}$, any $r > 0$, and any $x \in M$, $Vol_g(B_x(r)) \geq v(r)$. As a consequence, for any $(M,g) \in \mathcal{M}$ and any $\epsilon > 0$, the maximal number of disjoint balls of radius ϵ that M can contained is bounded above by $N = [V/v(\epsilon)] + 1$. In particular, there exists $d > 0$ such that for any $(M,g) \in \mathcal{M}$, $diam_{(M,g)} \leq d$ where $diam_{(M,g)}$ is the diameter of (M,g). Proposition 3.10 is then an easy consequence of proposition 5.2 of Gromov-Lafontaine-Pansu [GrLP]. This ends the proof of the proposition.

Let us now say some words about the $C^{k,\alpha}$ convergence of a sequence of compact Riemannian manifolds. Let n, k be integers, $\alpha \in (0,1)$, (M_j, g_j) a sequence of compact Riemannian n-manifolds, M a compact manifold, and g a $C^{k,\alpha}$ Riemannian metric on M. By definition (M_j, g_j) is said to be $C^{k,\alpha}$ convergent to (M,g) if there exists j_0 such that the following holds: for any $j \geq j_0$ there exist $C^{k+1,\alpha}$ diffeomorphisms $\Phi_j : M \to M_j$ such that in any chart of some subatlas of the C^∞ complete atlas of M, the components of the metrics $\Phi_j^* g_j$ converge $C^{k,\alpha}$ to the components of g. A set $\mathcal{M}$ of compact Riemannian n-manifolds is then said to be precompact in the $C^{k,\alpha}$-topology if any sequence in $\mathcal{M}$ possesses a $C^{k,\alpha}$ convergent subsequence. We refer the reader to the survey of Hebey-Herzlich [HH] for more details on these questions. For $n \geq 3$ an integer, $q \in [1,n)$, $\Lambda > 0$, $V > 0$, and $A > 0$ we now define $\tilde{\mathcal{M}}(n,q,\Lambda,V,A)$ as the set of compact Riemannian n-manifolds (M,g) such that:

(i) $|K_{(M,g)}| \leq \Lambda$ and $Vol_{(M,g)} \leq V$, $K_{(M,g)}$ the sectional curvature of (M,g)

(ii) for any $u \in C^\infty(M)$,

$$\left(\int_M |u|^{nq/(n-q)} dv(g) \right)^{(n-q)/nq} \leq A\left(\left(\int_M |\nabla u|^q dv(g) \right)^{1/q} + \left(\int_M |u|^q dv(g) \right)^{1/q} \right)$$

One then has the following.

Proposition 3.11: *For any n, $q \in [1,n)$, $\Lambda > 0$, $V > 0$, and $A > 0$, $\tilde{\mathcal{M}}(n,q,\Lambda,V,A)$ is precompact in the $C^{1,\alpha}$-topology for any $\alpha \in (0,1)$.*

Proof of proposition 3.11: As before there exist $v' > 0$ and $d > 0$ such that for any $(M,g) \in \tilde{\mathcal{M}}$, $Vol_{(M,g)} \geq v'$ and $diam_{(M,g)} \leq d$. Hence, by Cheeger-Gromov-Taylor [ChGT], there exist $i > 0$ such that for any $(M,g) \in \tilde{\mathcal{M}}$, $inj_{(M,g)} \geq i$.

Proposition 3.11 is then an easy consequence of theorem 1.1 of Anderson [An2] (see also [HH]). This ends the proof of the proposition.

By contradiction one then easily gets the following result, a kind of Cheeger's finitness theorem for $\tilde{\mathcal{M}}$.

Corollary 3.12: *For any n, $q \in [1, n)$, $\Lambda > 0$, $V > 0$, and $A > 0$, there are only finitely many diffeomorphism types of manifolds in $\tilde{\mathcal{M}}(n, q, \Lambda, V, A)$. In other words: there exists a finite number N of smooth compact manifolds $M_1, ..., M_N$ such that if $(M, g) \in \tilde{\mathcal{M}}$ then M is (smoothly) diffeomorphic to one of the M_i's.*

Remark: As a consequence of corollary 3.12, one easily gets that for any n, $q \in [1, n)$, $\Lambda > 0$, and $A > 0$ given, there are only finitely many diffeomorphism types of compact Riemannian n-manifolds (M, g) satisfying

$$|K_{(M,g)}|Vol_{(M,g)}^{2/n} \leq \Lambda$$

and such that for any $u \in C^\infty(M)$,

$$\left(\int_M |u|^{nq/(n-q)} dv(g) \right)^{(n-q)/nq}$$
$$\leq A \left(\left(\int_M |\nabla u|^q dv(g) \right)^{1/q} + Vol_{(M,g)}^{-1/n} \left(\int_M |u|^q dv(g) \right)^{1/q} \right)$$

One just has to note that the condition $|K_{(M,g)}|Vol_{(M,g)}^{2/n} \leq \Lambda$ and the Sobolev inequality above are scale invariant.

3.5 SOBOLEV EMBEDDINGS FOR COMPLETE MANIFOLDS

We discuss in this paragraph the existence of Sobolev embeddings for complete manifolds. We restrict ourselves to the study of the embeddings $H_1^q \subset L^p$, but recall that by lemma 3.1, if $H_1^1 \subset L^{n/(n-1)}$ then all the embeddings $H_k^q \subset H_m^p$ are valid. First we will see that, contrary to the case of compact manifolds, there exist complete manifolds for which the Sobolev embeddings are false. The existence of such manifolds is based on lemma 3.2. On the other hand, we will see in theorem 3.18 that the question of the existence of Sobolev embeddings for complete manifolds with Ricci curvature bounded from below is completely settled. For such manifolds the Sobolev embeddings are valid if and only if there is a uniform lower bound with respect to the center point x for the volume of balls $B_x(r)$, $r > 0$ arbitrary. As a consequence, we get by lemma 3.2 that if one of the embeddings $H_1^q \subset L^p$ is valid, $1 \leq q < n$ and $1/p = 1/q - 1/n$, then all the others are also valid. We refer to this property by saying that the scale of the embeddings $H_1^q \subset L^p$ is coherent.

As already mentioned, we first prove the following.

Proposition 3.13: *For any integer $n \geq 3$, there exist complete n-manifolds (M, g) for which the whole scale of the Sobolev embeddings $H_1^q \subset L^p$ is false, namely for which for any $1 \leq q < n$ and $1/p = 1/q - 1/n$, $H_1^q(M) \not\subset L^p(M)$.*

Proof of proposition 3.13: Consider the warped product

$$M = \mathbf{R} \times S^{n-1} \ , \ g(x, \theta) = \xi_x + u(x)h_\theta$$

where ξ is the euclidean metric of $\mathbf{R}$, h is the standard metric of the unit sphere S^{n-1} of $\mathbf{R}^n$, and $u : \mathbf{R} \to (0, 1]$ is smooth and such that $u(x) = 1$ when $x \leq 0$, $u(x) = e^{-2x}$ when $x \geq 1$. Clearly, if $y = (x_1, \theta_1)$ and $z = (x_2, \theta_2)$ are two points of M, then $d_g(y, z) \geq |x_2 - x_1|$. Hence, (M, g) is complete. In addition, if $y = (x, \theta)$ is a point of $M = \mathbf{R} \times S^{n-1}$, then $B_y(1) \subset (x - 1, x + 1) \times S^{n-1}$. As a consequence, when $x \geq 2$,

$$Vol_g\big(B_{(x,\theta)}(1)\big) \leq Vol_g\big((x - 1, x + 1) \times S^{n-1}\big)$$

$$\leq \omega_{n-1} \int_{x-1}^{x+1} e^{-(n-1)t}\, dt$$

$$\leq C(n)e^{-(n-1)x}$$

where ω_{n-1} denotes the volume of (S^{n-1}, h) and $C(n) = \frac{\omega_{n-1}}{(n-1)}\big(e^{n-1} - e^{1-n}\big)$. Therefore, for any $\theta \in S^{n-1}$,

$$\lim_{x \to \infty} Vol_g\big(B_{(x,\theta)}(1)\big) = 0$$

and by lemma 3.2 we get that $H_1^q(M) \not\subset L^p(M)$ for any $1 \leq q < n$ real and $1/p = 1/q - 1/n$. This ends the proof of the proposition.

Remark: The Ricci curvature of the manifold (M, g) constructed in the proof of proposition 3.13 is bounded from below. Indeed, since

$$Rc_{(S^{n-1}, h)} = (n - 2)h$$

one easily sees that $Rc_{(M,g)}$ will be bounded from below if there exists A real such that for any $\theta \in S^{n-1}$ and any $x > 1$, $Rc_{(M,g)}(x, \theta) \geq Ag_{(x,\theta)}$. Let

$$g'_{(x,\theta)} = e^{2x}\xi_x + h_\theta \ ,$$

R'_{ij} be the components of $Rc_{(M,g')}$ in some product chart $(\mathbf{R} \times \Omega, Id \times \phi)$, and R_{ij} be the components of $Rc_{(M,g)}$ in the same chart. We have $R'_{ij} = 0$ if $i = 1$ or $j = 1$, while $R'_{ij} = (n-2)h_{ij}$ if $i \geq 2$ and $j \geq 2$. Independently, if $g' = e^f g$ are conformal metrics on a n-manifold, then

$$R'_{ij} = R_{ij} - \frac{n-2}{2}(\nabla^2 f)_{ij} + \frac{n-2}{4}(\nabla f)_i (\nabla f)_j - \frac{1}{2}\left(-\Delta_g f + \frac{n-2}{2}|\nabla f|^2\right)g_{ij}$$

Hence, since $g' = e^{2x}g$ if $x > 1$, we get that for $x > 1$,

$$R_{11} = -(n-1) \text{ and } R_{1j} = 0 \text{ when } j \geq 2$$

$$R_{ij} = \left((n-2)e^{2x} - 1\right)g_{ij} \text{ when } i \geq 2 \text{ and } j \geq 2$$

As a consequence, $Rc_{(M,g)} \geq -(n-1)g$ for $x > 1$, and the Ricci curvature of (M,g) is bounded from below.

Let us now discuss the following result. This result was first proved by Varopoulos [Va1]. (We refer to paragraph 3.6 for the exact setting of Varopoulos result). The proof of Varopoulos was based on rather intricate semi-group technics. The proof we present here is more direct. It has its origins in Coulhon and Saloff-Coste [CoS].

Theorem 3.14: *Let (M,g) be a complete Riemannian n-manifold. Suppose that the Ricci curvature of (M,g) is bounded from below and that*

$$inf_{x \in M} Vol_g(B_x(1)) > 0$$

Then the Sobolev embeddings are valid for M.

Before we start with the proof of theorem 3.14, note that by Gromov's result the assumption $inf_{x \in M} Vol_g(B_x(1)) > 0$ implies that for any $r > 0$ there exists a positive constant v_r such that for any $x \in M$, $Vol_g(B_x(r)) \geq v_r$. (See theorem 1.5). Recall also that by lemma 3.1 we just have to prove that $H_1^1(M) \subset L^{n/(n-1)}(M)$.

The first step of the proof of theorem 3.14 we present is the following lemma of Coulhon and Saloff-Coste [CoS, section 3].

Lemma 3.15: *Let (M,g) be a complete Riemannian n-manifold such that $Rc_{(M,g)} \geq kg$ for some $k \in \mathbf{R}$, and let $R > 0$. There exists a positive constant $C = C(n,k,R)$, depending only on n, k, and R, such that for any $r \in (0,R)$, and any $u \in \mathcal{D}(M)$,*

$$\int_M |u - \bar{u}_r| dv(g) \leq Cr \int_M |\nabla u| dv(g)$$

where $\bar{u}_r(x) = \frac{1}{Vol_g(B_x(r))} \int_{B_x(r)} u\,dv(g),\ x \in M$.

Proof of lemma 3.15: Let (M, g) be a complete Riemannian n-manifold such that $Rc_{(M,g)} \geq kg$ for some $k \in \mathbf{R}$, and let $R > 0$. By the work of Buser [Bu], there exists a positive constant $C = C(n, k, R)$, depending only on n, k, and R, such that for any $x \in M$, any $r \in (0, 2R)$, and any $u \in C^\infty(B_x(r))$,

$$\int_{B_x(r)} |u - \bar{u}_r(x)|dv(g) \leq Cr \int_{B_x(r)} |\nabla u|dv(g) \qquad (5)$$

Let $r \in (0, R)$ be given and let $(x_i)_{i \in I}$ be a sequence of points of M such that simultaneously $M = \cup_i B_{x_i}(2r)$ and $B_{x_i}(r) \cap B_{x_j}(r) = \emptyset$ when $i \neq j$. If we proceed as in the proof of lemma 1.6 we get that

$$\mathrm{Card}\{i \in I \text{ s.t. } x \in B_{x_i}(2r)\} \leq N = N(n, k, R) = (16)^n e^{8\sqrt{(n-1)|k|}R}$$

Let $u \in \mathcal{D}(M)$. We have

$$\int_M |u - \bar{u}_r|dv(g) \leq \sum_i \int_{B_{x_i}(r)} |u - \bar{u}_r|dv(g)$$

$$\leq \sum_i \int_{B_{x_i}(r)} |u - \bar{u}_r(x_i)|dv(g)$$

$$+ \sum_i \int_{B_{x_i}(r)} |\bar{u}_r(x_i) - \bar{u}_{2r}(x_i)|dv(g)$$

$$+ \sum_i \int_{B_{x_i}(r)} |\bar{u}_r - \bar{u}_{2r}(x_i)|dv(g)$$

By (5), we get that

$$\sum_i \int_{B_{x_i}(r)} |u - \bar{u}_r(x_i)|dv(g) \leq Cr \sum_i \int_{B_{x_i}(r)} |\nabla u|dv(g)$$

$$\leq NCr \int_M |\nabla u|dv(g)$$

while

$$\sum_i \int_{B_{x_i}(r)} |\bar{u}_r(x_i) - \bar{u}_{2r}(x_i)|dv(g) = \sum_i Vol_g(B_{x_i}(r))|\bar{u}_r(x_i) - \bar{u}_{2r}(x_i)|$$

$$\leq \sum_i \int_{B_{x_i}(r)} |u - \bar{u}_{2r}(x_i)|dv(g)$$

$$\leq \sum_i \int_{B_{x_i}(2r)} |u - \bar{u}_{2r}(x_i)|dv(g)$$

$$\leq 2NCr \int_M |\nabla u|dv(g)$$

Independently, we have

$$\sum_i \int_{B_{x_i}(r)} |\bar{u}_r - \bar{u}_{2r}(x_i)| dv(g)$$

$$\leq \sum_i \int_{x \in B_{x_i}(r)} \left\{ \frac{1}{Vol_g(B_x(r))} \int_{y \in B_x(r)} |u(y) - \bar{u}_{2r}(x_i)| dv_g(y) \right\} dv_g(x)$$

$$\leq \sum_i \int_{x \in B_{x_i}(r)} \left\{ \frac{1}{Vol_g(B_x(r))} \int_{y \in B_{x_i}(2r)} |u(y) - \bar{u}_{2r}(x_i)| dv_g(y) \right\} dv_g(x)$$

$$\leq \sum_i \int_{B_{x_i}(2r)} |u(y) - \bar{u}_{2r}(x_i)| dv_g(y) \int_{B_{x_i}(r)} \frac{1}{Vol_g(B_x(r))} dv_g(x)$$

But, by (5),

$$\int_{B_{x_i}(2r)} |u(y) - \bar{u}_{2r}(x_i)| dv_g(y) \leq 2Cr \int_{B_{x_i}(2r)} |\nabla u| dv(g)$$

while, by Gromov's result,

$$\frac{1}{Vol_g(B_x(r))} \leq \frac{K}{Vol_g(B_x(2r))}$$

where $K = K(n, k, R) = 2^n e^{2\sqrt{(n-1)|k|}R}$. Since $x \in B_{x_i}(r)$ implies that $B_{x_i}(r)$ is a subset of $B_x(2r)$, we get that

$$\int_{B_{x_i}(r)} \frac{1}{Vol_g(B_x(r))} dv_g(x) \leq K$$

Hence,

$$\sum_i \int_{B_{x_i}(r)} |\bar{u}_r - \bar{u}_{2r}(x_i)| dv(g) \leq 2KCNr \int_M |\nabla u| dv(g)$$

and for any $u \in \mathcal{D}(M)$,

$$\int_M |u - \bar{u}_r| dv(g) \leq 3(1 + K)NCr \int_M |\nabla u| dv(g)$$

This ends the proof of the lemma.

Let us now prove the following.

Lemma 3.16: *Let (M, g) be a complete Riemannian n-manifold. Suppose that its Ricci curvature satisfies $Rc_{(M,g)} \geq kg$ for some $k \in \mathbf{R}$ and suppose that there exists $v > 0$ such that $Vol_g(B_x(1)) \geq v$ for any $x \in M$. There exist two positive constants $C = C(n, k, v)$ and $\eta = \eta(n, k, v)$, depending only on n, k, and v, such* *that for any open subset Ω of M with smooth boundary and compact closure, if $Vol_g(\Omega) \leq \eta$, then $Vol_g(\Omega)^{(n-1)/n} \leq C\,Area_g(\partial\Omega)$.*

Proof of lemma 3.16: By theorem 1.5 and the remark following this theorem we have that for any $x \in M$ and any $0 < r < R$,

$$Vol_g\left(B_x(r)\right) \geq \left\{\frac{1}{R^n}e^{-\sqrt{(n-1)|k|}R}Vol_g\left(B_x(R)\right)\right\} r^n$$

Fix $R = 1$. Then we get that for any $x \in M$ and any $r \in (0,1)$,

$$Vol_g\left(B_x(r)\right) \geq \left(e^{-\sqrt{(n-1)|k|}}v\right) r^n$$

Set $\eta = \frac{1}{16}e^{-\sqrt{(n-1)|k|}}v$ and $C_1 = e^{-\sqrt{(n-1)|k|}}v$. Let Ω be some open subset of M with smooth boundary, compact closure, and such that $Vol_g(\Omega) \leq \eta$. For sufficiently small $\epsilon > 0$ consider the function

$u_\epsilon(x) = 1$ if $x \in \Omega$

$u_\epsilon(x) = 1 - \frac{1}{\epsilon}d_g(x,\partial\Omega)$ if $x \in M\backslash\Omega$ and $d_g(x,\partial\Omega) \leq \epsilon$

$u_\epsilon(x) = 0$ if $x \in M\backslash\Omega$ and $d_g(x,\partial\Omega) \geq \epsilon$.

Then u_ϵ is lipschitz for every ϵ and one easily sees that

$$\lim_{\epsilon \to 0}\int_M u_\epsilon\,dv(g) = Vol_g(\Omega)$$

and

$|\nabla u_\epsilon| = \frac{1}{\epsilon}$ if $x \in M\backslash\overline{\Omega}$ and $d_g(x,\partial\Omega) < \epsilon$

$|\nabla u_\epsilon| = 0$ otherwise

which implies that

$$\lim_{\epsilon \to 0}\int_M |\nabla u_\epsilon|\,dv(g) = Area_g(\partial\Omega)$$

Furthermore, for every $\epsilon > 0$, $Vol_g(\Omega) = Vol_g\left(\{x \in M \text{ s.t. } u_\epsilon(x) \geq 1\}\right)$, and for any $\epsilon > 0$ and any $r > 0$,

$$Vol_g\left(\{x \in M \text{ s.t. } u_\epsilon(x) \geq 1\}\right) \leq Vol_g\left(\{x \in M \text{ s.t. } |u_\epsilon(x) - \overline{u}_{\epsilon,r}(x)| \geq \frac{1}{2}\}\right)$$

$$+ Vol_g\left(\{x \in M \text{ s.t. } \overline{u}_{\epsilon,r}(x) \geq \frac{1}{2}\}\right)$$

where $\overline{u}_{\epsilon,r}(x) = \frac{1}{Vol_g\left(B_x(r)\right)}\int_{B_x(r)} u_\epsilon\,dv(g)$. Now, note that for $r > 0$ and $\epsilon \ll 1$,

$$\overline{u}_{\epsilon,r}(x) \leq \frac{2Vol_g(\Omega)}{Vol_g\left(B_x(r)\right)}$$

Fix $r = \left(\frac{8Vol_g(\Omega)}{C_1}\right)^{1/n}$. Since $Vol_g(\Omega) \le \eta = \frac{C_1}{16}$, we get that $r \in (0,1)$ and that

$$\frac{2Volg(\Omega)}{Vol_g(B_x(r))} \le \frac{1}{4}$$

(according to what we have said above). Hence,

$$\left\{x \in M \text{ s.t. } \overline{u}_{\epsilon,r}(x) \ge \frac{1}{2}\right\} = \emptyset$$

and for every $0 < \epsilon << 1$,

$$Vol_g(\Omega) \le Vol_g\left(\left\{x \in M \text{ s.t. } |u_\epsilon(x) - \overline{u}_{\epsilon,r}(x)| \ge \frac{1}{2}\right\}\right)$$

But

$$Vol_g\left(\left\{x \in M \text{ s.t. } |u_\epsilon(x) - \overline{u}_{\epsilon,r}(x)| \ge \frac{1}{2}\right\}\right) \le 2\int_M |u_\epsilon - \overline{u}_{\epsilon,r}|dv(g)$$

and by lemma 3.15 there exists a positive constant $C_2 = C_2(n,k)$ such that

$$\int_M |u_\epsilon - \overline{u}_{\epsilon,r}|dv(g) \le C_2 r \int_M |\nabla u_\epsilon|dv(g)$$

Hence,

$$Vol_g(\Omega) \le 2C_2\left(\frac{8Vol_g(\Omega)}{C_1}\right)^{1/n} \lim_{\epsilon \to 0} \int_M |\nabla u_\epsilon|dv(g)$$
$$\le C_3 Vol_g(\Omega)^{1/n} \cdot Area_g(\partial\Omega)$$

where C_3 depends only on n, k, and v. This ends the proof of the lemma.

It is by now classical (Federer-Fleming [Fe], [FeF], see also Chavel [Ch]) that as a consequence of lemma 3.16 one has the following.

Lemma 3.17: *Let (M,g) be a complete Riemannian n-manifold. Suppose that its Ricci curvature satisfies $Rc_{(M,g)} \ge kg$ for some $k \in \mathbf{R}$ and suppose that there exists $v > 0$ such that $Vol_g(B_x(1)) \ge v$ for any $x \in M$. There exist two positive constants $\delta = \delta(n,k,v)$ and $A = A(n,k,v)$, depending only on n, k, and v, such that for any $x \in M$ and any $u \in \mathcal{D}(B_x(\delta))$,*

$$\left(\int_M |u|^{n/(n-1)}dv(g)\right)^{(n-1)/n} \le A\int_M |\nabla u|dv(g)$$

Proof of lemma 3.17: Let $\eta = \eta(n, k, v)$ be as in lemma 3.16. By theorem 1.5 there exists $\delta = \delta(n, k, v)$ such that for any $x \in M$, $Vol_g(B_x(\delta)) \leq \eta$. Let $x \in M$ and let $u \in \mathcal{D}(B_x(\delta))$. For $t \geq 0$, let

$$\Omega(t) = \{x \in M \text{ s.t. } |u(x)| > t\} \quad \text{and} \quad V(t) = Vol_g(\Omega(t))$$

Clearly, $V(t) \leq \eta$ for any $t \geq 0$. Then the coarea formula and lemma 3.16 imply that

$$\int_M |\nabla u| dv(g) \geq \frac{1}{C} \int_0^\infty V(t)^{1-1/n} dt$$

where C is the constant given by lemma 3.16. Independently,

$$\int_M |u|^{n/(n-1)} dv(g) = \frac{n}{n-1} \int_0^\infty t^{1/(n-1)} V(t) dt$$

Noting that

$$\int_0^\infty V(t)^{1-1/n} dt \geq \left(\frac{n}{n-1} \int_0^\infty t^{1/(n-1)} V(t) dt \right)^{1-1/n}$$

this ends the proof of the lemma. For more details, we refer to Chavel [Ch, chapter 6].

With such a result we are now in position to prove theorem 3.14.

Proof of theorem 3.14: Let (M, g) be a complete Riemannian n-manifold such that $Rc_{(M,g)} \geq kg$ for some $k \in \mathbf{R}$ and such that there exists $v > 0$ with $Vol_g(B_x(1)) \geq v$ for any $x \in M$. We want to prove that the Sobolev embeddings are valid for M. As already mentioned, by lemma 3.1 we just have to prove that $H_1^1(M) \subset L^{n/(n-1)}(M)$.

Let $\delta = \delta(n, k, v)$ be as in lemma 3.17 and (x_i) be a sequence of points of M such that

(1) $M = \cup_i B_{x_i}(\delta/2)$ and $B_{x_i}(\delta/4) \cap B_{x_j}(\delta/4) = \emptyset$ if $i \neq j$

(2) there exists $N = N(n, k, v)$, depending only on n, k, and v such that each point of M has a neighborhood which intersects at most N of the $B_{x_i}(\delta)$'s.

The existence of such a sequence is given by lemma 1.6. Let $\rho : [0, \infty) \to [0, 1]$ be defined by

$\rho(t) = 1$ if $0 \leq t \leq \delta/2$

$\rho(t) = 3 - (4/\delta)t$ if $\delta/2 \leq t \leq 3\delta/4$

$\rho(t) = 0$ if $t \geq 3\delta/4$

and let $\alpha_i(x) = \rho\big(d_g(x_i, x)\big)$, $x \in M$. Clearly α_i is lipschitz with compact support. Hence, see lemma 2.5, $\alpha_i \in H_1^1(M)$. Furthermore, since we have that $Supp\,\alpha_i \subset B_{x_i}(3\delta/4)$, we get without any difficulty that $\alpha_i \in \overset{\circ}{H}{}_1^1(B_{x_i}(\delta))$. Let $\eta_i = \alpha_i / \sum_m \alpha_m$. Then, since $|\nabla\alpha_i| \leq 4/\delta$ a.e., we get by (2) that for any i, $\eta_i \in \overset{\circ}{H}{}_1^1(B_{x_i}(\delta))$, (η_i) is a partition of unity subordinate to the covering $(B_{x_i}(\delta))$, $\nabla\eta_i$ exists almost everywhere, and there exists a positive constant $H = H(n, k, v)$ such that $|\nabla\eta_i| \leq H$ a.e.

Let $u \in \mathcal{D}(M)$. We have

$$\left(\int_M |u|^{n/(n-1)} dv(g)\right)^{(n-1)/n} \leq \sum_i \left(\int_M |\eta_i u|^{n/(n-1)} dv(g)\right)^{(n-1)/n}$$

$$\leq A \sum_i \int_M |\nabla(\eta_i u)| dv(g)$$

where A is the constant of lemma 3.17. Hence,

$$\left(\int_M |u|^{n/(n-1)} dv(g)\right)^{(n-1)/n} \leq A \sum_i \int_M \eta_i |\nabla u| dv(g) + A \sum_i \int_M |u||\nabla\eta_i| dv(g)$$

$$\leq A \int_M |\nabla u| dv(g) + ANH \int_M |u| dv(g)$$

$$\leq A(1 + NH)\left\{ \int_M |\nabla u| dv(g) + \int_M |u| dv(g)\right\}$$

and there exists $A' > 0$ such that for any $u \in \mathcal{D}(M)$,

$$\left(\int_M |u|^{n/(n-1)} dv(g)\right)^{(n-1)/n} \leq A'\left\{ \int_M |\nabla u| dv(g) + \int_M |u| dv(g)\right\}$$

By theorem 2.7 we then get that $H_1^1(M) \subset L^{n/(n-1)}(M)$. As already mentioned, this ends the proof of the theorem.

Combining theorem 3.14 with lemma 3.2 one easily gets what we announced at the beginning of this paragraph. Namely, the scale of the embeddings $H_1^q \subset L^p$ is coherent for complete manifolds with Ricci curvature bounded from below, and the Sobolev embeddings are valid if and only if one has a uniform lower bound with respect to the center point x for the volume of balls $B_x(r)$, $r > 0$ arbitrary. More precisely, we have the following.

Theorem 3.18: *Let (M, g) be a complete Riemannian n-manifold with Ricci curvature bounded from below.*

(1) If for some $1 \leq q < n$ and $1/p = 1/q - 1/n$, $H_1^q(M) \subset L^p(M)$, then for any $1 \leq q < n$ and $1/p = 1/q - 1/n$, $H_1^q(M) \subset L^p(M)$. In particular, $H_1^1(M) \subset L^{n/(n-1)}(M)$.

(2) The Sobolev embeddings are valid for M if and only if there is a uniform lower bound for the volume of balls which is independent of their center, namely if and only if $\inf_{x \in M} Vol_g(B_x(1)) > 0$.

By Croke's result [Cr1, proposition 14], we get as an easy consequence of theorem 3.18 the following corollary. As a remark, note that Aubin [Au1] and Cantor [Can] proved in the middle of the '70's that the Sobolev embeddings were valid for complete manifolds with bounded sectional curvature and positive injectivity radius.

Corollary 3.19: *The Sobolev embeddings are valid for complete manifolds with Ricci curvature bounded from below and positive injectivity radius.*

Remark: Let (M, g) be a complete Riemannian n-manifold. Suppose that its Ricci curvature satisfies $Rc_{(M,g)} \geq kg$ for some $k \in \mathbf{R}$ and suppose that there exists $v > 0$ such that $Vol_g(B_x(1)) \geq v$ for any $x \in M$. According to what we have said above, for any $1 \leq q < n$ and $1/p = 1/q - 1/n$, there exists a positive constant A such that for any $u \in H_1^q(M)$,

$$\left(\int_M |u|^p dv(g) \right)^{1/p} \leq A \left\{ \left(\int_M |\nabla u|^q dv(g) \right)^{1/q} + \left(\int_M |u|^q dv(g) \right)^{1/q} \right\}$$

The proof of theorem 3.14 and the remark following lemma 3.1 give the exact dependence of A. Namely, A depends only on n, q, k, and v.

We have seen in paragraph 3.3 that for compact manifolds $H_1^q \subset L^p$ for any $1 \leq q < n$ and any $p \geq 1$ such that $p \leq nq/(n-q)$. (i.e $1/p \geq 1/q - 1/n$). We prove below that such a result is still valid for complete manifolds satisfying the assumptions of theorem 3.14, provided one also has that $p \geq q$. Note here that for $1 \leq p < q$, the embedding of H_1^q in L^p is not valid in general. Think for instance to $\mathbf{R}^n$ endowed with the euclidean metric, and let $u_\alpha \in C^\infty(\mathbf{R}^n)$ be such that $u_\alpha(x) = 1/|x|^\alpha$ if $|x| \geq 1$. One easily checks that for $1 \leq p < q$, $u_{n/p} \in H_1^q(\mathbf{R}^n)$ while $u_{n/p} \notin L^p(\mathbf{R}^n)$.

Theorem 3.20: *Let (M, g) be a complete Riemannian n-manifold with Ricci curvature bounded from below and such that $\inf_{x \in M} Vol_g(B_x(1)) > 0$. For any $1 \leq q < n$ and any p such that $q \leq p \leq nq/(n-q)$, $H_1^q(M) \subset L^p(M)$.*

Proof of theorem 3.20: Let $k \in \mathbf{R}$ be such that $Rc_{(M,g)} \geq kg$, and let $v > 0$ be such that for any $x \in M$, $Vol_g(B_x(1)) \geq v$. Fix $1 \leq q < n$ and let $p' = nq/(n-q)$. By theorem 3.18 there exists a positive constant $A = A(n, q, k, v)$ such that for any $u \in H_1^q(M)$,

$$\left(\int_M |u|^{p'} dv(g) \right)^{1/p'} \leq A \left\{ \left(\int_M |\nabla u|^q dv(g) \right)^{1/q} + \left(\int_M |u|^q dv(g) \right)^{1/q} \right\}$$

Independently, by theorem 1.5 there exists a positive constant $V = V(n, k)$ such that for any $x \in M$, $Vol_g(B_x(4)) \leq V$. Combining these two results we get that for any $1 \leq p \leq nq/(n-q)$, any $x \in M$, and any $u \in \mathcal{D}(B_x(4))$,

$$\left(\int_{B_x(4)} |u|^p dv(g) \right)^{1/p} \leq AV^{1/p - 1/p'} \left\{ \left(\int_{B_x(4)} |\nabla u|^q dv(g) \right)^{1/q} \right.$$

$$\left. + \left(\int_{B_x(4)} |u|^q dv(g) \right)^{1/q} \right\}$$

Let (x_i) and (α_i) be as in the proof of theorem 3.14 with $\delta = 4$. Set $\eta_i = \alpha_i^{[q]+1} / \sum_m \alpha_m^{[q]+1}$. Then for any i, $\eta_i^{1/q} \in \overset{\circ}{H}_1^q(B_{x_i}(4))$, (η_i) is a partition of unity subordinate to the covering $(B_{x_i}(4))$, $\nabla \eta_i^{1/q}$ exists almost everywhere, and there exists a positive constant $H = H(n, q, k, v)$ such that $|\nabla \eta_i^{1/q}| \leq H$ a.e.

Let $u \in \mathcal{D}(M)$ and let p be such that $q \leq p \leq nq/(n-q)$. Then, since $p/q \geq 1$,

$$\|u\|_p^q = \|u^q\|_{p/q} = \|\sum_i \eta_i u^q\|_{p/q} \leq \sum_i \|\eta_i u^q\|_{p/q} \leq \sum_i \|\eta_i^{1/q} u\|_p^q$$

where $\|u\|_p = \left(\int_M |u|^p dv(g) \right)^{1/p}$. Hence, according to what we have said above,

$$\left(\int_M |u|^p dv(g) \right)^{q/p} \leq (AV^{1/p - 1/p'})^q \mu \left\{ \sum_i \int_M |\nabla(\eta_i^{1/q} u)|^q dv(g) \right.$$

$$\left. + \sum_i \int_M \eta_i |u|^q dv(g) \right\}$$

where $\mu = \mu(q)$ is such that for any $x, y \geq 0$, $(x+y)^q \leq \mu(x^q + y^q)$.

Set $C = (AV^{1/p - 1/p'})^q \mu$. We then get that

$$\left(\int_M |u|^p dv(g) \right)^{q/p} \leq C \int_M |u|^q dv(g)$$

$$+ C \sum_i \int_M (|\nabla u| \eta_i^{1/q} + |u| |\nabla \eta_i^{1/q}|)^q dv(g)$$

$$\leq C \int_M |u|^q dv(g) + C\mu \sum_i \int_M |\nabla u|^q \eta_i dv(g)$$

$$+ C\mu \sum_i \int_M |\nabla \eta_i^{1/q}|^q |u|^q dv(g)$$

$$\leq C\mu \int_M |\nabla u|^q dv(g) + C(1 + \mu N H^q) \int_M |u|^q dv(g)$$

where $N = N(n,k)$ is such that each point of M has a neighborhood which intersects at most N of the $B_{x_i}(4)$'s. Finally, since $(x+y)^{1/q} \leq x^{1/q} + y^{1/q}$ for any $x, y \geq 0$, we get that for any $u \in \mathcal{D}(M)$,

$$\left(\int_M |u|^p dv(g) \right)^{1/p} \leq A' \left\{ \left(\int_M |\nabla u|^q dv(g) \right)^{1/q} + \left(\int_M |u|^q dv(g) \right)^{1/q} \right\}$$

where $A' = C^{1/q} \max(\mu, 1 + \mu N H^q)^{1/q}$. By theorem 2.7, this ends the proof of the theorem.

Remarks: 1) Here again we know the exact dependence of the constant A of the embedding of $H_1^q(M)$ in $L^p(M)$. Namely, A depends on n, p, q, a lower bound for the Ricci curvature of (M, g), and the constant v such that $Vol_g(B_x(1)) \geq v$ for any $x \in M$.

2) The assumption on the Ricci curvature we made till now is satisfactory but not necessary. Indeed, there exist complete manifolds for which the embeddings of H_1^q in L^p are valid for all $1 \leq q < n$ and $1/p = 1/q - 1/n$, but for which the Ricci curvature is not bounded from below. Think for instance to $\mathbf{R}^n$ endowed with the metric

$$g_{ij} = e^{sin(\pi r^2/2)} \delta_{ij}$$

where $r = |x|$. One clearly has that $(1/e)\delta_{ij} \leq g_{ij} \leq e\delta_{ij}$ as bilinear forms. Hence, by theorem 3.4, one easily sees that the embeddings $H_1^q \subset L^p$ are valid for such a metric. Independently, it is easy to see that the Ricci curvature of g is not bounded from below. Recall for that purpose that if $g' = e^f g$ are conformal metrics on a n-manifold, then

$$R'_{ij} = R_{ij} - \frac{n-2}{2}(\nabla^2 f)_{ij} + \frac{n-2}{4}(\nabla f)_i(\nabla f)_j - \frac{1}{2}\left(-\Delta_g f + \frac{n-2}{2}|\nabla f|^2\right)g_{ij}$$

where the R_{ij}'s and R'_{ij}'s denote the components of $Rc_{(M,g)}$ and $Rc_{(M,g')}$.

3) We have seen in paragraph 3.3 that for compact manifolds, the embeddings $H_1^q \subset L^p$ are compact for $p < nq/(n-q)$. This is clearly false for complete manifolds. Just think to $\mathbf{R}^n$ endowed with the euclidean metric e, and let

$u \in \mathcal{D}(\mathbf{R}^n)$ be such that $0 \le u \le 1$, $u = 1$ in $B_0^e(1)$, and $u = 0$ in $\mathbf{R}^n \backslash B_0^e(2)$. For m integer, set $u_m(x) = u(x - x_m)$ where $x_m \in \mathbf{R}^n$ is such that $|x_m| = m$. Clearly, (u_m) is bounded in $H_1^q(\mathbf{R}^n)$ by $\|u\|_{H_1^q}$, while for any m, $\|u_m\|_p = \|u\|_p > 0$. As a consequence, and since (u_m) converges to 0 for the pointwise convergence, it does not converge in $L^p(\mathbf{R}^n)$. On the other hand, as we will see in paragraph 5.3, symmetries can help to reverse the situation.

3.6 DISTURBED INEQUALITIES FOR COMPLETE MANIFOLDS

As already mentioned, theorem 3.14 is a corollary of Varopoulos result [Val, section 0.3]. The exact setting of this result is that for any complete Riemannian n-manifold (M, g) satisfying $Rc_{(M,g)} \ge kg$ for some $k \in \mathbf{R}$, there exists a positive constant $A = A(n, k)$, depending only on n and k, such that for any $u \in \mathcal{D}(M)$,

$$\left(\int_M |u|^{n/(n-1)} v \, dv(g) \right)^{(n-1)/n} \le A \int_M \left(|\nabla u| + |u| \right) v \, dv(g)$$

where $v(x) = \dfrac{1}{Vol_g(B_x(1))}$, $x \in M$. (The assumptions of theorem 3.14 imply that v is bounded above while by theorem 1.5, v is bounded from below. It is then easy to obtain theorem 3.14 from such an inequality). The proof of Varopoulos was based on rather intricate semi-group technics. Independently, Maheux and Saloff-Coste [MaS] obtained by direct arguments the following result which, in a certain sense, generalizes the one of Buser [Bu] to Sobolev inequalities. We refer to [MaS] for its proof.

Theorem 3.21: *Let (M, g) be a complete Riemannian n-manifold such that $Rc_{(M,g)} \ge kg$ for some $k \in \mathbf{R}$, and let p, q be two real numbers such that $1 \le q < n$ and $p \in [q, nq/(n-q)]$. There exist $C = C_n > 0$ and $A = A_{p,q} > 0$ such that for any $x \in M$, any $r > 0$, and any $u \in C^\infty(B_x(r))$,*

$$\left(\int_{B_x(r)} |u - \bar{u}_r(x)|^p dv(g) \right)^{1/p}$$

$$\le A e^{C(1+\sqrt{|k|}r)} r \, Vol_g(B_x(r))^{1/p - 1/q} \left(\int_{B_x(r)} |\nabla u|^q dv(g) \right)^{1/q}$$

where $\bar{u}_r(x) = \dfrac{1}{Vol_g(B_x(r))} \int_{B_x(r)} u \, dv(g)$.

For $q = 1$ and $p = n/(n-1)$ we get in particular that for any $r > 0$ there exists a positive constant $A = A(n, k, r)$, depending only on n, k, and r, such that for any $x \in M$ and any $u \in C^\infty(B_x(r))$,

$$\left(\int_{B_x(r)} |u|^p dv(g) \right)^{1/p}$$

$$\leq A Vol_g \left(B_x(r) \right)^{-1/n} \left(\int_{B_x(r)} |\nabla u| dv(g) + \int_{B_x(r)} |u| dv(g) \right) \qquad (\star)_r$$

With such inequalities we then get the following extensions of Varopoulos result (proposition 3.22 and theorem 3.24). In particular, this provides us with a simple and direct proof of this result.

Proposition 3.22: *Let (M, g) be a complete Riemannian n-manifold. Suppose that $Rc_{(M,g)} \geq kg$ for some $k \in \mathbf{R}$. Then for any $(\alpha, \beta) \in \mathbf{R} \times \mathbf{R}$ satisfying $\beta - \frac{n-1}{n}\alpha \geq 1/n$ there exists a positive constant $A = A(n, k, \alpha, \beta)$, depending only on n, k, α, and β, such that for any $u \in \mathcal{D}(M)$,*

$$\left(\int_M |u|^{n/(n-1)} v^\alpha dv(g) \right)^{(n-1)/n} \leq A \int_M \left(|\nabla u| + |u| \right) v^\beta dv(g)$$

where $v(x) = \dfrac{1}{Vol_g(B_x(1))}$, $x \in M$.

Proof of proposition 3.22: Let (x_i) be a sequence of points of M such that

(1) $M = \cup_i B_{x_i}(1/2)$ and $B_{x_i}(1/4) \cap B_{x_j}(1/4) = \emptyset$ if $i \neq j$

(2) there exists $N = N(n, k)$, depending only on n and k, such that each point of M has a neighborhood which intersects at most N of the $B_{x_i}(1)$'s.

The existence of such a sequence is given by lemma 1.6. Let $\rho : [0, \infty) \to [0, 1]$ be defined by

$\rho(t) = 1$ if $0 \leq t \leq 1/2$

$\rho(t) = 3 - 4t$ if $1/2 \leq t \leq 3/4$

$\rho(t) = 0$ if $t \geq 3/4$

and let $\alpha_i(x) = \rho(d_g(x_i, x))$, $x \in M$. Clearly α_i is lipschitz with compact support. Hence, see lemma 2.5, $\alpha_i \in H_1^1(M)$. Furthermore, since we have that $Supp \alpha_i \subset B_{x_i}(3/4)$, we get without any difficulty that $\alpha_i \in \overset{\circ}{H}_1^1(B_{x_i}(1))$. Let $\eta_i = \alpha_i / \sum_m \alpha_m$. Then, since $|\nabla \alpha_i| \leq 4$ a.e., we get by (2) that for any i, $\eta_i \in \overset{\circ}{H}_1^1(B_{x_i}(1))$, (η_i) is a partition of unity subordinate to the covering $(B_{x_i}(1))$, $\nabla \eta_i$ exists almost everywhere, and there exists a positive constant $H = H(N)$ such that $|\nabla \eta_i| \leq H$ a.e.

Independently, by theorem 1.5 we get that there exists a positive constant $C = C(n,k)$ such that for any $x \in M$, $Vol_g(B_x(1)) \geq C Vol_g(B_x(2))$. As a consequence, for any $x \in M$ and any $y \in B_x(1)$, $Vol_g(B_x(1)) \geq C Vol_g(B_y(1))$ (since $y \in B_x(1)$ implies that $B_y(1) \subset B_x(2)$). Similarly, for any $x \in M$ and any $y \in B_x(1)$, we clearly have (by symmetry) that $Vol_g(B_y(1)) \geq C Vol_g(B_x(1))$. Furthermore, again by theorem 1.5, there exists a positive constant $V = V(n,k)$ such that for any $x \in M$, $Vol_g(B_x(1)) \leq V$.

Let $\alpha, \beta \in \mathbf{R}$ be such that $\beta - \frac{n-1}{n}\alpha \geq \frac{1}{n}$. Multiplying $(\star)_1$ by

$$Vol_g(B_x(1))^{-(n-1)\alpha/n},$$

and according to what we have just said, we get that there exists a positive constant $A' = A'(n,k)$ such that for any $x \in M$ and any $u \in C^{\infty}(B_x(1))$,

$$\left(\int_{B_x(1)} |u|^{n/(n-1)} v^{\alpha} dv(g)\right)^{(n-1)/n}$$

$$\leq A' Vol_g(B_x(1))^{-((n-1)\alpha+1)/n} \int_{B_x(1)} \left(|\nabla u| + |u|\right) dv(g)$$

$$\leq A' Vol_g(B_x(1))^{\beta - ((n-1)\alpha+1)/n} Vol_g(B_x(1))^{-\beta} \int_{B_x(1)} \left(|\nabla u| + |u|\right) dv(g)$$

$$\leq A'' \int_{B_x(1)} \left(|\nabla u| + |u|\right) v^{\beta} dv(g)$$

where $A'' = A' V^{\beta - ((n-1)\alpha+1)/n} C^{-|\beta|}$ depends only on n, k, α, and β. As a consequence, for any i and any $u \in C^{\infty}(B_{x_i}(1))$,

$$\left(\int_{B_{x_i}(1)} |u|^{n/(n-1)} v^{\alpha} dv(g)\right)^{(n-1)/n} \leq A'' \int_{B_{x_i}(1)} \left(|\nabla u| + |u|\right) v^{\beta} dv(g)$$

Let $u \in \mathcal{D}(M)$. We then have

$$\left(\int_M |u|^{n/(n-1)} v^{\alpha} dv(g)\right)^{(n-1)/n}$$

$$\leq \sum_i \left(\int_{B_{x_i}(1)} |\eta_i u|^{n/(n-1)} v^{\alpha} dv(g)\right)^{(n-1)/n}$$

$$\leq A'' \sum_i \int_M \left(|\nabla(\eta_i u)| + \eta_i |u|\right) v^{\beta} dv(g)$$

$$\leq A'' \sum_i \int_M |\nabla \eta_i| |u| v^{\beta} dv(g) + A'' \sum_i \int_M \eta_i |\nabla u| v^{\beta} dv(g)$$

$$+ A'' \int_M |u| v^\beta \, dv(g)$$

$$\leq A'' N H \int_M |u| v^\beta \, dv(g) + A'' \int_M |\nabla u| v^\beta \, dv(g)$$

$$+ A'' \int_M |u| v^\beta \, dv(g)$$

$$\leq A \int_M \Big(|\nabla u| + |u| \Big) v^\beta \, dv(g)$$

where $A = A''(1 + NH)$ depends only on n, k, α, and β. This ends the proof of the proposition.

Corollary 3.23: *Let (M, g) be a complete Riemannian n-manifold such that $Rc_{(M,g)} \geq kg$ for some $k \in \mathbf{R}$. There exists a positive constant $A = A(n, k)$, depending only on n and k, such that for any $u \in \mathcal{D}(M)$,*

$$\left(\int_M |u|^{n/(n-1)} v \, dv(g) \right)^{(n-1)/n} \leq A \int_M \Big(|\nabla u| + |u| \Big) v \, dv(g)$$

(Varopoulos inequality) and

$$\left(\int_M |u|^{n/(n-1)} dv(g) \right)^{(n-1)/n} \leq A \int_M \Big(|\nabla u| + |u| \Big) v^{1/n} \, dv(g)$$

where $v(x) = \dfrac{1}{Vol_g\big(B_x(1)\big)}$, $x \in M$.

Proof of corollary 3.23: Just take $\alpha = \beta = 1$ (resp. $\alpha = 0$ and $\beta = 1/n$) in the inequality of proposition 3.22.

Theorem 3.24: *Let (M, g) be a complete Riemannian n-manifold such that $Rc_{(M,g)} \geq kg$ for some $k \in \mathbf{R}$, and let (p, q) be two real numbers such that $1 \leq q < n$ and $1/p = 1/q - 1/n$. Then for any $(\alpha, \beta) \in \mathbf{R} \times \mathbf{R}$ satisfying $\beta/q - \alpha/p \geq 1/n$, there exists a positive constant $A = A(n, q, k, \alpha, \beta)$, depending only on n, q, k, α, and β, such that for any $u \in \mathcal{D}(M)$,*

$$\left(\int_M |u|^p v^\alpha \, dv(g) \right)^{1/p} \leq A \left(\left(\int_M |\nabla u|^q v^\beta \, dv(g) \right)^{1/q} + \left(\int_M |u|^q v^\beta \, dv(g) \right)^{1/q} \right)$$

where $v(x) = \dfrac{1}{Vol_g\big(B_x(1)\big)}$, $x \in M$.

Proof of theorem 3.24: We proceed as in the proof of lemma 3.1, but starting from proposition 3.22. Let α, β be as in proposition 3.22, and let (p, q) be two real numbers such that $1 < q < n$ and $1/p = 1/q - 1/n$. By proposition 3.22 there exists $A = A(n, k, \alpha, \beta) > 0$ such that for any $u \in \mathcal{D}(M)$,

$$\left(\int_M |u|^{n/(n-1)} v^\alpha dv(g) \right)^{(n-1)/n} \leq A \int_M \left(|\nabla u| + |u| \right) v^\beta dv(g)$$

Let $u \in \mathcal{D}(M)$ and set $\phi = |u|^{p(n-1)/n}$. Applying Hölder's inequality we get that

$$\left(\int_M |u|^p v^\alpha dv(g) \right)^{(n-1)/n}$$

$$= \left(\int_M |\phi|^{n/(n-1)} v^\alpha dv(g) \right)^{(n-1)/n}$$

$$\leq A \int_M \left(|\nabla \phi| + |\phi| \right) v^\beta dv(g)$$

$$\leq \frac{Ap(n-1)}{n} \int_M |u|^{p'} |\nabla u| v^\beta dv(g)$$

$$\quad + A \int_M |u|^{p(n-1)/n} v^\beta dv(g)$$

$$\leq \frac{Ap(n-1)}{n} \left(\int_M |u|^{p'q'} v^\alpha dv(g) \right)^{1/q'} \left(\int_M |\nabla u|^q v^{q(\beta-\alpha/q')} dv(g) \right)^{1/q}$$

$$\quad + A \left(\int_M |u|^{p'q'} v^\alpha dv(g) \right)^{1/q'} \left(\int_M |u|^q v^{q(\beta-\alpha/q')} dv(g) \right)^{1/q}$$

where $1/q + 1/q' = 1$ and $p' = p(n-1)/n - 1$. But $p'q' = p$ since $1/p = 1/q - 1/n$. As a consequence, for any $u \in \mathcal{D}(M)$,

$$\left(\int_M |u|^p v^\alpha dv(g) \right)^{1/p}$$

$$\leq \frac{Ap(n-1)}{n} \left(\left(\int_M |\nabla u|^q v^\gamma dv(g) \right)^{1/q} + \left(\int_M |u|^q v^\gamma dv(g) \right)^{1/q} \right)$$

where $\gamma = q(\beta - \alpha/q')$. Noting that $\gamma/q - \alpha/p = \beta - \frac{n-1}{n}\alpha$, this ends the proof of the theorem.

Remark: Varopoulos inequalities combined with Gromov's result furnish the inequalities $(\star)_r$ for any $x \in M$, any $r > 0$, and any $u \in \mathcal{D}(B_x(r))$. About such inequalities $(\star)_r$, see also Saloff-Coste [Sa1, theorem 10.3].

Let us now say some words about the following result of Schoen-Yau [ScY]. Here again, the norms will be disturbed by some function. Recall that if $(\tilde{M}, \tilde{g})$ and (M, g) are Riemannian manifolds, an immersion $\phi : (\tilde{M}, \tilde{g}) \to (M, g)$ is said to be conformal if there exists a smooth function $f : \tilde{M} \to \mathbf{R}$ such that $\phi^\star g = e^f \tilde{g}$. In the following, (S^n, h) denotes the standard unit sphere of $\mathbf{R}^{n+1}$.

Proposition 3.25: *Let (M,g) be a Riemannian n-manifold (not necessarily complete). Assume that there exists a conformal immersion*

$$\phi : (M,g) \rightarrow (S^n, h)$$

For any $u \in \mathcal{D}(M)$,

$$\left(\int_M |u|^{2n/(n-2)} dv(g) \right)^{(n-2)/n}$$

$$\leq \frac{4}{n(n-2)\omega_n^{2/n}} \left(\int_M |\nabla u|^2 dv(g) + \frac{n-2}{4(n-1)} \int_M Scal_{(M,g)} u^2 dv(g) \right)$$

where ω_n is the volume of (S^n, h) and $Scal_{(M,g)}$ is the scalar curvature of (M,g).

Proof of proposition 3.25: Define

$$Q(M) = inf_{\{u \in \mathcal{D}(M), \int_M |u|^{2n/(n-2)} dv(g)=1\}} \int_M u(L_g u) dv(g)$$

where

$$L_g u = \Delta_g u + \frac{n-2}{4(n-1)} Scal_{(M,g)} u$$

is the conformal laplacian. One easily checks that for any $v \in C^\infty(M)$, $v > 0$, and any $u \in C^\infty(M)$,

$$L_g(uv) = v^{(n+2)/(n-2)} L_{g'}(u)$$

where $g' = v^{4/(n-2)}g$. By Obata [Ob], $Q(S^n) = (n(n-2)\omega_n^{2/n})/4$. The inequality of proposition 3.25 is then equivalent to $Q(M) \geq Q(S^n)$. Let $(\Omega_i)_{i \in \mathbf{N}}$ be an exhaustion of M by compact domains with smooth boundary. We then have

$$Q(M) = \lim_{i \to \infty} Q(\Omega_i)$$

Thus, in order to show that $Q(M) \geq Q(S^n)$, it is enough to show that $Q(\Omega) \geq Q(S^n)$ for any domain $\Omega \subset M$ with $\overline{\Omega}$ compact and $\partial\Omega$ smooth. Now, the proof is by contradiction. Hence, we suppose that for some Ω as above, $Q(\Omega) < Q(S^n)$. By standard variational techniques one then gets that there exists a smooth function $u > 0$ in Ω satisfying $\int_\Omega u^{2n/(n-2)} dv(g) = 1$ as well as

$$L_g u = Q(\Omega) u^{(n+2)/(n-2)} \text{ in } \Omega, \quad u = 0 \text{ on } \partial\Omega$$

If we extend u by defining $u \equiv 0$ in $M \backslash \Omega$ we then have

$$L_g u \leq Q(\Omega) u^{(n+2)/(n-2)} \text{ in } M, \quad \int_M u^{2n/(n-2)} dv(g) = 1$$

where the inequality is understood in the distributional sense. Let ϕ be the conformal immersion of proposition 3.25, $\phi : (M, g) \rightarrow (S^n, h)$. We define a function $\tilde{u}$ on S^n by $\tilde{u} \equiv 0$ in $S^n \backslash \phi(\overline{\Omega})$, and for $y \in \phi(\overline{\Omega})$,

$$\tilde{u}(y) = max_{x \in \phi^{-1}(y) \cap \overline{\Omega}} \; \alpha(x)^{-(n-2)/4} u(x)$$

where $\phi^\star h = \alpha g$. Since ϕ is an immersion, the set $\phi^{-1}(y) \cap \overline{\Omega}$ is finite, and for each $x \in \phi^{-1}(y) \cap \overline{\Omega}$ there is a neighborhood U_x of x such that ϕ is a diffeomorphism of U_x onto $\phi(U_x)$, a neighborhood of y. Let ϕ_x^{-1} denote the inverse of this local diffeomorphism. By conformal invariance property of the conformal laplacian, if

$$\tilde{u}_x(y) = \beta_x(y)^{(n-2)/4} u\big(\phi_x^{-1}(y)\big), \; y \in \phi(U_x)$$

where $(\phi_x^{-1})^\star g = \beta_x h$, then $L_h \tilde{u}_x \leq Q(\Omega) \tilde{u}_x^{(n+2)/(n-2)}$ in $\phi(U_x)$. Hence, we see that $\tilde{u}$ is a nonnegative lipschitzian function on S^n satisfying

$$L_h \tilde{u} \leq Q(\Omega) \tilde{u}^{(n+2)/(n-2)}$$

on S^n. Again by conformal invariance we have

$$\int_{\phi(U_x)} \tilde{u}_x^{2n/(n-2)} dv(h) = \int_{U_x} u^{2n/(n-2)} dv(g)$$

and hence we see that $\int_{S^n} \tilde{u}^{2n/(n-2)} dv(h) \leq 1$. By integrating the differential inequality satisfied by $\tilde{u}$ one then gets that

$$\int_{S^n} \tilde{u}(L_h \tilde{u}) dv(h) \leq Q(\Omega) \int_{S^n} \tilde{u}^{2n/(n-2)} dv(h)$$

Since $\int_{S^n} \tilde{u}^{2n/(n-2)} dv(h) \leq 1$, this inequality implies that $Q(S^n) \leq Q(\Omega)$, a contradiction. This ends the proof of proposition 3.25.

Combining proposition 3.25 with the fact that for any simply connected conformally flat Riemannian n-manifold (M, g) there exists a conformal immersion Φ from (M, g) to (S^n, h), see for instance Kulkarni in [KuP], one gets corollary 3.26. Recall that (M, g) is said to be conformally flat if for each point $x \in M$ there exists a chart (Ω, ϕ) of M at x, and a smooth function $f : \Omega \rightarrow \mathbf{R}$, such that $(\phi^{-1})^\star (e^f g)_{ij} = \delta_{ij}$ for any $i, j = 1 \ldots, n$. Spaces of constant sectional curvature are conformally flat. More generally, when $n \geq 4$, (M, g) is conformally flat if and only if its Weyl curvature vanishes identically.

Corollary 3.26: *Let* (M, g) *be a complete conformally flat Riemannian n-manifold. For any simply connected domain Ω of M and any $u \in \mathcal{D}(\Omega)$,*

$$\left(\int_M |u|^{2n/(n-2)} dv(g) \right)^{(n-2)/n}$$
$$\leq \frac{4}{n(n-2)\omega_n^{2/n}} \left(\int_M |\nabla u|^2 dv(g) + \frac{n-2}{4(n-1)} \int_M Scal_{(M,g)} u^2 dv(g) \right)$$

Remark: Let (M, g) be a compact Riemannian n-manifold. Suppose g is Einstein. By Obata [Ob] one has that if (M, g) is not conformally diffeomorphic to (S^n, h) then, up to a constant scale factor, g is the unique metric of constant scalar curvature in its conformal class. As a consequence of the resolution of the Yamabe problem by Aubin and Schoen (see the historical notes at the end of paragraph 4.2) one then easily gets that for any compact Einstein n-manifold (M, g), any g' in the conformal class of g, and any $u \in C^\infty(M)$,

$$\frac{n-2}{4(n-1)} Scal_{(M,g)} Vol_{(M,g)}^{2/n} \left(\int_M |u|^{2n/(n-2)} dv(g') \right)^{(n-2)/n}$$
$$\leq \int_M |\nabla u|^2 dv(g') + \frac{n-2}{4(n-1)} \int_M Scal_{(M,g')} u^2 dv(g')$$

When $Scal_{(M,g)} > 0$, this provides us with (disturbed) Sobolev inequalities. Such inequalities can be useful. We refer the reader to Hebey-Vaugon [HV4] for an example of application.

3.7 $\mathbf{R}^n$-TYPE INEQUALITIES FOR COMPLETE MANIFOLDS

Let (M, g) be a complete Riemannian n-manifold of infinite volume and let $q \in [1, n)$. In view of theorem 3.4, one can ask under what conditions there exists $C_q > 0$ such that for any $u \in \mathcal{D}(M)$,

$$\left(\int_M |u|^{nq/(n-q)} dv(g) \right)^{(n-q)/n} \leq C_q \int_M |\nabla u|^q dv(g)$$

(The manifold constructed in the proof of proposition 3.13 provides us with an example of complete manifolds with infinite volume such that for any $q \in [1, n)$, C_q does not exist). We will mainly be concerned here with the case $q = 2$. However, note that similar arguments to those developed in the proof of lemma 3.1 prove that if C_2 exists, then C_q exists for any $2 \leq q < n$. Indeed, let $u \in \mathcal{D}(M)$ and set $\phi = |u|^{p(n-2)/2n}$ where $p = nq/(n-q)$, $q \in [2, n)$. If C_2

exists, applying Hölder's inequality we get that

$$\left(\int_M |u|^p dv(g)\right)^{(n-2)/n}$$

$$= \left(\int_M |\phi|^{2n/(n-2)} dv(g)\right)^{(n-2)/n}$$

$$\leq C_2 \int_M |\nabla \phi|^2 dv(g)$$

$$= (p'+1)^2 C_2 \int_M |u|^{2p'} |\nabla u|^2 dv(g)$$

$$\leq (p'+1)^2 C_2 \left(\int_M |u|^{2p'q/(q-2)} dv(g)\right)^{(q-2)/q} \left(\int_M |\nabla u|^q dv(g)\right)^{2/q}$$

where $p' = \frac{p(n-2)}{2n} - 1$. But $2p'q/(q-2) = p$ since $1/p = 1/q - 1/n$. As a consequence, we get that for any $q \in [2, n)$ and any $u \in \mathcal{D}(M)$,

$$\left(\int_M |u|^{nq/(n-q)} dv(g)\right)^{(n-q)/n} \leq \left(\frac{q(n-2)}{2(n-q)}\right)^q C_2^{q/2} \int_M |\nabla u|^q dv(g)$$

This proves the claim. In the following, if Ω is a regular bounded open subset of M (we will write $\Omega \subset\subset M$), $\lambda_1^D(\Omega)$ denotes the first eigenvalue of the laplacian for the Dirichlet problem on Ω. One then has the following result of Carron [Car2].

Proposition 3.27: *Let (M, g) be a complete Riemannian n-manifold of infinite volume. The following two propositions are equivalent.*

(1) There exists $C_2 > 0$ such that for any $u \in \mathcal{D}(M)$,

$$\left(\int_M |u|^{2n/(n-2)} dv(g)\right)^{(n-2)/n} \leq C_2 \int_M |\nabla u|^2 dv(g)$$

(2) There exists $\Lambda > 0$ such that for any $\Omega \subset\subset M$, $\lambda_1^D(\Omega) \geq \Lambda Vol_g(\Omega)^{-2/n}$.

Proof of proposition 3.27: The proof follows the lines of [Car2]. Suppose first that there exists $C_2 > 0$ such that for any $u \in \mathcal{D}(M)$,

$$\left(\int_M |u|^{2n/(n-2)} dv(g)\right)^{(n-2)/n} \leq C_2 \int_M |\nabla u|^2 dv(g)$$

By Hölder's inequality one easily gets that if Ω is some regular bounded open subset of M and if $u \not\equiv 0$ satisfies $\Delta_g u = \lambda_1^D(\Omega)u$ in Ω, $u = 0$ on $\partial\Omega$, then

$$\frac{\int_\Omega |\nabla u|^2 dv(g)}{\left(\int_\Omega |u|^{2n/(n-2)} dv(g)\right)^{(n-2)/n}} \leq \lambda_1^D(\Omega) Vol_g(\Omega)^{2/n}$$

Hence, for any regular bounded open subset Ω of M, $\lambda_1^p(\Omega) \geq C_2^{-1} Vol_g(\Omega)^{-2/n}$. This proves that (1) implies (2). Suppose now that there exists $\Lambda > 0$ such that for any $\Omega \subset\subset M$, $\lambda_1^p(\Omega) \geq \Lambda Vol_g(\Omega)^{-2/n}$. Let $\Omega \subset\subset M$ be given. For $p > n$ we set

$$\Lambda_p(\Omega) = inf_{U \subset\subset \Omega} \lambda_1^p(U) Vol_g(U)^{2/p}$$

$$\mu_p(\Omega) = inf_{u \in \mathcal{D}(\Omega)} \frac{\int_M |\nabla u|^2 dv(g)}{\left(\int_M |u|^{2p/(p-2)} dv(g)\right)^{(p-2)/p}}$$

Since $2p/(p-2) < 2n/(n-2)$ when $p > n$, by standard arguments one easily obtains that for any $p > n$ there exists $u_p \in C^\infty(\Omega) \cap \overset{\circ}{H}_1^2(\Omega)$ such that

$$\Delta_g u_p = \mu_p(\Omega) u_p^{(p+2)/(p-2)} \text{ in } \Omega, \ u_p > 0 \text{ in } \Omega, \ \int_\Omega u_p^{2p/(p-2)} dv(g) = 1$$

One can then prove (see [Car2]) that for any $0 \leq t \leq \|u_p\|_\infty$,

$$Vol_g\left(\{x \in \Omega \text{ s.t. } u_p(x) > \|u_p\|_\infty - t\}\right)$$

$$\geq \left(\frac{\Lambda_p(\Omega)}{2^{(p+4)/4}}\right)^{p/2} \left(\frac{t}{\mu_p(\Omega)\|u_p\|_\infty^{(p+2)/(p-2)}}\right)^{p/2}$$

Hence, if we set $L = \|u_p\|_\infty$, we get that

$$1 = \int_\Omega u_p^{2p/(p-2)} dv(g)$$

$$= \frac{2p}{(p-2)} \int_0^L Vol_g\left(\{x \in \Omega \text{ s.t. } u_p(x) > t\}\right) t^{(p+2)/(p-2)} dt$$

$$= \frac{2p}{(p-2)} \int_0^L Vol_g\left(\{x \in \Omega \text{ s.t. } u_p(x) > L - t\}\right)(L-t)^{(p+2)/(p-2)} dt$$

$$\geq \frac{2p}{(p-2)} \left(\frac{\Lambda_p(\Omega)}{2^{(p+4)/4}\mu_p(\Omega)L^{(p+2)/(p-2)}}\right)^{p/2} \int_0^L t^{p/2}(L-t)^{(p+2)/(p-2)} dt$$

But

$$\int_0^1 \theta^{x-1}(1-\theta)^{y-1} d\theta = \frac{\Gamma(x)\Gamma(y)}{\Gamma(x+y)}$$

where $\Gamma(x) = \int_0^\infty t^{x-1}e^{-t} dt$ is the Euler function, and

$$\int_0^L t^{p/2}(L-t)^{(p+2)/(p-2)} dt = L^{p(p+2)/2(p-2)} \int_0^1 \theta^{p/2}(1-\theta)^{(p+2)/(p-2)} d\theta$$

As a consequence, we get that for any $p > n$,

$$\Lambda_p(\Omega) \leq C(p)\mu_p(\Omega)$$

where

$$C(p) = 2^{(p+4)/4}\left(\frac{2p\Gamma(1+p/2)\Gamma(2p/(p-2))}{(p-2)\Gamma((p^2+4p-4)/2(p-2))}\right)^{-2/p}$$

Now, by hypothesis,

$$\Lambda_p(\Omega) \geq \Lambda Vol_g(\Omega)^{2/p-2/n}$$

Hence, if $u \in \mathcal{D}(\Omega)$,

$$\left(\int_M |u|^{2n/(n-2)}dv(g)\right)^{(n-2)/n}$$

$$= \lim_{p \to n^+}\left(\int_M |u|^{2p/(p-2)}dv(g)\right)^{(p-2)/p}$$

$$\leq \lim_{p \to n^+}\left(C(p)\Lambda_p(\Omega)^{-1}\right)\int_M |\nabla u|^2 dv(g)$$

$$\leq \frac{1}{\Lambda}\left(\lim_{p \to n^+} C(p)Vol_g(\Omega)^{2/n-2/p}\right)\int_M |\nabla u|^2 dv(g)$$

$$\leq \frac{C(n)}{\Lambda}\int_M |\nabla u|^2 dv(g)$$

and we get that for any $\Omega \subset\subset M$ and any $u \in \mathcal{D}(\Omega)$,

$$\left(\int_M |u|^{2n/(n-2)}dv(g)\right)^{(n-2)/n} \leq \frac{C(n)}{\Lambda}\int_M |\nabla u|^2 dv(g)$$

This ends the proof of the proposition.

Remark: Inequalities such as $\lambda_1^p(\Omega) \geq \Lambda Vol_g(\Omega)^{-2/n}$ for any $\Omega \subset\subset M$, are referred to as Faber-Krahn's inequalities.

Let us now say some words about the existence of positive Green's functions on complete non-compact Riemannian manifolds. Let (M, g) be a complete non-compact Riemannian manifold and let x be some point of M. One can then prove that, uniformly with respect to x, either there exist positive Green's functions of pole x, and in particular there exists a positive minimal Green's function of pole x, or there does not exist any positive Green's function of pole x. More precisely, let $\Omega \subset\subset M$ be such that $x \in \Omega$ and let G be the solution of

$$\begin{cases} \Delta_g G = \delta_x \text{ in } \Omega \\ G = 0 \text{ on } \partial\Omega \end{cases}$$

Set $G_x^\Omega(y) = G(y)$ when $y \in \Omega$, $G_x^\Omega(y) = 0$ otherwise. Obviously, $G_x^\Omega \leq G_x^{\Omega'}$ if $\Omega \subset \Omega'$.

One then has the following.

Theorem 3.28: *Set $G_x(y) = sup_{\{\Omega \ s.t. \ x \in \Omega\}} G_x^\Omega(y)$, $y \in M$. Then,*

(i) *either $G_x(y) = +\infty$, $\forall y \in M$*

(ii) *or $G_x(y) < +\infty$, $\forall y \in M \backslash \{x\}$.*

This alternative does not depend on x and in case (ii), G_x is the positive minimal Green's function of pole x.

In case (i) the manifold is said to be parabolic, in case (ii) the manifold is said to be non-parabolic. By Cheng-Yau [ChY], one has that if for some $x \in M$,

$$\liminf_{r \to +\infty} \frac{Vol_g(B_x(r))}{r^2} < +\infty$$

then (M, g) is parabolic. This explains for instance why $\mathbf{R}^2$ is parabolic while $\mathbf{R}^3$ is not. More generally, it is proved in Grigor'yan [Gr2] and Varopoulos [Va2] that if for some $x \in M$,

$$\int_1^{+\infty} \frac{rdr}{Vol_g(B_x(r))} = +\infty$$

then, again, (M, g) is parabolic. Conversely, Varopoulos proved in [Va2] that if the Ricci curvature of (M, g) is nonnegative and if

$$\int_1^{+\infty} \frac{rdr}{Vol_g(B_x(r))} < +\infty$$

then (M, g) is non-parabolic. Independently, see Grigor'yan [Gr2], one has that if

$$\int_1^{+\infty} \frac{dr}{h(r)^2} < +\infty$$

where

$$h(r) = inf_{\{\Omega \subset\subset M \ s.t. \ Vol_g(\Omega) \leq r\}} Area_g(\partial\Omega)$$

then (M, g) is non-parabolic. For more details on these questions we refer the reader to Cheng-Yau [ChY], Grigor'yan [Gr2], Varopoulos [Va2], and the references contained in these papers.

Let us now prove the following result. It is extracted from Carron [Car2].

Theorem 3.29: *Let (M, g) be a complete Riemannian n-manifold of infinite volume. The following two propositions are equivalent.*

(1) There exists $C_2 > 0$ such that for any $u \in \mathcal{D}(M)$,

$$\left(\int_M |u|^{2n/(n-2)} dv(g) \right)^{(n-2)/n} \leq C_2 \int_M |\nabla u|^2 dv(g)$$

(2) (M, g) is non-parabolic and there exists $K > 0$ such that for any $x \in M$ and any $t > 0$,

$$Vol_g\left(\{ y \in M \text{ s.t. } G_x(y) > t \} \right) \leq K t^{-n/(n-2)}$$

where G_x is the positive minimal Green's function of pole x.

Proof of theorem 3.29: The proof follows the lines of [Car2]. Suppose first that there exists $C_2 > 0$ such that for any $u \in \mathcal{D}(M)$,

$$\left(\int_M |u|^{2n/(n-2)} dv(g) \right)^{(n-2)/n} \leq C_2 \int_M |\nabla u|^2 dv(g)$$

Let $x \in M$, let Ω be some regular bounded open subset of M such that $x \in \Omega$, and set

$$\Phi_t(y) = min\left(G_x^{\Omega}(y), t \right)$$

where G_x^{Ω} is as in theorem 3.28 and where $t > 0$ is given. Applying the inequality above to Φ_t we get that

$$\int_M |\nabla \Phi_t|^2 dv(g) \geq C_2^{-1} \left(\int_M \Phi_t^{2n/(n-2)} dv(g) \right)^{(n-2)/n}$$

$$\geq C_2^{-1} Vol_g\left(\{ y \in M \text{ s.t. } G_x^{\Omega}(y) > t \} \right)^{(n-2)/n} t^2$$

while if $\Theta = \{ y \in M \text{ s.t. } G_x^{\Omega}(y) < t \}$ and $\theta = \{ y \in M \text{ s.t. } G_x^{\Omega}(y) = t \}$,

$$\int_M |\nabla \Phi_t|^2 dv(g) = \int_{\Theta} |\nabla G_x^{\Omega}|^2 dv(g)$$

$$= \int_{\Theta} G_x^{\Omega}(\Delta_g G_x^{\Omega}) dv(g) - t \int_{\theta} (\partial_n G_x^{\Omega}) ds$$

$$= t$$

since $\Delta_g G_x^{\Omega} = 0$ in $\Omega \backslash \{x\}$. As a consequence, for any $x \in M$, any $t > 0$, and any bounded open subset Ω of M such that $x \in \Omega$,

$$Vol_g\left(\{ y \in M \text{ s.t. } G_x^{\Omega}(y) > t \} \right) \leq C_2^{n/(n-2)} t^{-n/(n-2)}$$

By theorem 3.28 one then gets that (M, g) is non-parabolic and that for any $x \in M$ and any $t > 0$,

$$Vol_g\left(\{ y \in M \text{ s.t. } G_x(y) > t \} \right) \leq C_2^{n/(n-2)} t^{-n/(n-2)}$$

where G_x is the positive minimal Green's function of pole x. This proves that (1) implies (2).

Suppose now that (M, g) is non-parabolic and that there exists $K > 0$ such that for any $x \in M$ and any $t > 0$,

$$Vol_g\left(\{y \in M \text{ s.t. } G_x(y) > t\}\right) \le Kt^{-n/(n-2)}$$

Let Ω be some regular bounded open subset of M and let $u \not\equiv 0$ be such that

$$\Delta_g u = \lambda_1^D(\Omega)u \text{ in } \Omega, \ u = 0 \text{ on } \partial\Omega$$

For any $x \in \Omega$,

$$u(x) = \lambda_1^D(\Omega) \int_\Omega G_x^\Omega u \, dv(g)$$

$$\le \lambda_1^D(\Omega)\left(\int_\Omega G_x^\Omega dv(g)\right)\left(sup_{y \in \Omega} u(y)\right)$$

where G_x^Ω is as in theorem 3.28. We now choose $x \in \Omega$ such that

$$u(x) = sup_{y \in \Omega} u(y)$$

Since $G_x^\Omega \le G_x$, we get that

$$1 \le \lambda_1^D(\Omega)\int_\Omega G_x dv(g) = \lambda_1^D(\Omega)\int_0^{+\infty} Vol_g\left(\{y \in \Omega \text{ s.t. } G_x(y) > t\}\right)dt$$

while, by hypothesis,

$$Vol_g\left(\{y \in \Omega \text{ s.t. } G_x(y) > t\}\right) \le min\left(Vol_g(\Omega), Kt^{-n/(n-2)}\right)$$

As a consequence,

$$\int_0^{+\infty} Vol_g\left(\{y \in \Omega \text{ s.t. } G_x(y) > t\}\right)dt \le TVol_g(\Omega) + K\int_T^{+\infty} t^{-n/(n-2)}dt$$

where T is such that $KT^{-n/(n-2)} = Vol_g(\Omega)$. Hence,

$$\int_0^{+\infty} Vol_g\left(\{y \in \Omega \text{ s.t. } G_x(y) > t\}\right)dt \le \frac{n}{2}K^{(n-2)/n}Vol_g(\Omega)^{2/n}$$

and we get we get that for any $\Omega \subset\subset M$,

$$\lambda_1^D(\Omega) \ge \frac{2}{n}K^{-(n-2)/n}Vol_g(\Omega)^{-2/n}$$

By theorem 3.27, this ends the proof of the theorem.

Let us now say some words about the case $q = 1$. Namely, we ask if there exists $C_1 > 0$ such that for any $u \in \mathcal{D}(M)$,

$$\left(\int_M |u|^{n/(n-1)} dv(g) \right)^{(n-1)/n} \leq C_1 \int_M |\nabla u| dv(g)$$

By the work of Hoffman and Spruck [HoS], we know that C_1 exists if M is simply connected with nonpositive sectional curvature. (See also the remark after theorem 4.4). Independently, and since the existence of C_1 implies the existence of C_2 (see the proof of lemma 3.1), by theorem 3.29 we get that if C_1 exists then (M, g) is non-parabolic and there exists $K > 0$ such that for any $x \in M$ and any $t > 0$,

$$Vol_g \left(\{ y \in M \text{ s.t. } G_x(y) > t \} \right) \leq K t^{-n/(n-2)}$$

where G_x is the positive minimal Green's function of pole x. Such a result was already contained in Grigor'yan [Gr2]. One can now ask if this condition is also sufficient. The answer is yes if the Ricci curvature of the manifold is nonnegative, but the result is false in general. (See Carron [Car2], Coulhon-Ledoux [CoL], and Varopoulos [Va3]).

Proposition 3.30: *(1) Let (M, g) be a complete Riemannian manifold of infinite volume. If the Ricci curvature of (M, g) is nonnegative, the existence of C_1 is equivalent to the existence of C_2.*

(2) There exist complete Riemannian manifolds of infinite volume for which C_2 exists but C_1 does not exist. In addition, one can choose these manifolds such that the sectional curvature is bounded and the injectivity radius is positive.

Finally, see for instance [Car2] for more details, we mention that if (M, g) is a non-parabolic complete Riemannian n-manifold whose Ricci curvature is bounded from below, and if there exists $K > 0$ such that for any $x \in M$ and any $t > 0$ the positive minimal Green's function G_x of pole x satisfies

$$Vol_g \left(\{ y \in M \text{ s.t. } G_x(y) > t \} \right) \leq K t^{-n/(n-1)}$$

then C_1 exists. However, this result is not sharp. Indeed, C_1 exists for the Euclidean space $\mathbf{R}^n$ while the condition above is obviously not satisfied by the positive minimal Green's function

$$G_x(y) = \frac{1}{(n-2)\omega_{n-1}|y - x|^{n-2}}$$

of $\mathbf{R}^n$.

Remarks: 1) Similar results to those presented in this paragraph are available if for some $p \in (2, 2n/(n-2))$, one asks for the existence of $C_2^p > 0$ such that for any $u \in \mathcal{D}(M)$,

$$\left(\int_M |u|^p dv(g) \right)^{2/p} \leq C_2^p \int_M |\nabla u|^2 dv(g)$$

One has for instance that C_2^p exists if and only if (M, g) is non-parabolic and if there exists $K > 0$ such that for any $x \in M$ and any $t > 0$,

$$Vol_g \left(\{ y \in M \text{ s.t. } G_x(y) > t \} \right) \leq Kt^{-p/2}$$

where G_x is the positive minimal Green's function of pole x. We refer the reader to [Car2] for more details. Independently, we mention that Yau proved in [Ya] that if (M, g) is a complete simply connected n-manifold with sectional curvature less than $K < 0$, then its Cheeger's constant is greater than or equal to $(n-1)\sqrt{-K}$. For such manifolds one then has that for any $u \in \mathcal{D}(M)$,

$$\int_M |u| dv(g) \leq (n-1)\sqrt{-K} \int_M |\nabla u| dv(g)$$

Clearly, this implies that for any $p \geq 1$ and any $u \in \mathcal{D}(M)$,

$$\left(\int_M |u|^p dv(g) \right)^{1/p} \leq (n-1)p\sqrt{-K} \left(\int_M |\nabla u|^p dv(g) \right)^{1/p}$$

To see this, one can proceed as in the proof of lemma 3.1 by setting $\phi = |u|^p$ in Yau's inequality.

2) For complete manifolds of finite volume, the $\mathbf{R}^n$-type Sobolev inequalities are obviously false. One can then ask if the Sobolev-Poincaré inequalities of proposition 3.9 are valid for such manifolds. (This could be motivated by the idea that the $\mathbf{R}^n$-type inequalities are the infinite volume versions of the Sobolev-Poincaré inequalities). As it can easily be deduced from lemma 3.2, the answer to this question is no ...unless M is compact. Indeed, suppose that for some $1 \leq q < n$ and $1/p = 1/q - 1/n$, there exists $A > 0$ such that for any $u \in \mathcal{D}(M)$,

$$\left(\int_M |u - \overline{u}|^p dv(g) \right)^{1/p} \leq A \left(\int_M |\nabla u|^q dv(g) \right)^{1/q}$$

Then for any $u \in \mathcal{D}(M)$,

$$\left(\int_M |u|^p dv(g) \right)^{1/p} \leq A \left(\int_M |\nabla u|^q dv(g) \right)^{1/q} + Vol_{(M,g)}^{-1/n} \left(\int_M |u|^q dv(g) \right)^{1/q}$$

(see the proof of proposition 4.1), and by lemma 3.2 one gets that there exists some positive constant v such that for any $x \in M$, $Vol_g\big(B_x(1)\big) \geq v$. This implies that the maximal number of disjoint balls of radius 1 that M can contained is bounded above by $[Vol_{(M,g)}/v] + 1$. This in turn implies that M must be compact. The claim is proved.

Chapter 4

The best constants problems

Let (M, g) be a compact Riemannian n-manifold. By theorem 3.5 we know that for any real number $1 \leq q < n$, $H_1^q(M) \subset L^p(M)$ where $1/p = 1/q - 1/n$. From now on, we will write that for any $1 \leq q < n$ and $1/p = 1/q - 1/n$, there exist two real numbers A and B, depending a priori on the manifold and q, such that for any $u \in H_1^q(M)$,

$$\left(\int_M |u|^p dv(g) \right)^{1/p} \leq A \left(\int_M |\nabla u|^q dv(g) \right)^{1/q} + B \left(\int_M |u|^q dv(g) \right)^{1/q} \qquad (I_q)$$

We define

$$\mathcal{A}_q(M) = \{ A \in \mathbf{R} \text{ s.t. } B \text{ exists in } (I_q) \}$$

$$\mathcal{B}_q(M) = \{ B \in \mathbf{R} \text{ s.t. } A \text{ exists in } (I_q) \}$$

Clearly, if $A \in \mathcal{A}_q(M)$ (resp. $B \in \mathcal{B}_q(M)$) and if $A' \geq A$ (resp. $B' \geq B$), then $A' \in \mathcal{A}_q(M)$ (resp. $B' \in \mathcal{B}_q(M)$). We set $\alpha_q(M) = inf\mathcal{A}_q(M)$ and $\beta_q(M) = inf\mathcal{B}_q(M)$. By definition, $\alpha_q(M)$ and $\beta_q(M)$ are the best constants.

Two symmetrical research programs are associated to the inequalities (I_q). In the first one priority is given to the constant A, in the second one priority is given to the constant B. We state these two programs in the case of compact manifolds although program $\mathcal{A}$ will be considered in the more general setting of complete manifolds.

Program $\mathcal{A}$	Program $\mathcal{B}$
Question 1$\mathcal{A}$: is it possible to compute explicitly $\alpha_q(M)$?	Question 1$\mathcal{B}$: is it possible to compute explicitly $\beta_q(M)$?
Question 2$\mathcal{A}$: is $\mathcal{A}_q(M)$ a closed set ? Namely, does one have that $\alpha_q(M) \in \mathcal{A}_q(M)$?	Question 2$\mathcal{B}$: is $\mathcal{B}_q(M)$ a closed set ? Namely, does one have that $\beta_q(M) \in \mathcal{B}_q(M)$?
Question 3$\mathcal{A}$: for $A \in \mathcal{A}_q(M)$ close to $\alpha_q(M)$ (resp. for $A = \alpha_q(M)$ if $\mathcal{A}_q(M)$ is closed), on what quantities does the constant B of (I_q) depends ?	Question 3$\mathcal{B}$: for $B \in \mathcal{B}_q(M)$ close to $\beta_q(M)$ (resp. for $B = \beta_q(M)$ if $\mathcal{B}_q(M)$ is closed), on what quantities does the constant A of (I_q) depends ?
Question 4$\mathcal{A}$: can one compute explicitly the constant B of (I_q) for large classes of manifolds when $A = \alpha_q(M)$?	Question 4$\mathcal{B}$: can one compute explicitly the constant A of (I_q) for large classes of manifolds when $B = \beta_q(M)$?

Finally we will be concerned with the following question which establishes a connection between these two programs.

> Question 5: For which manifolds one has that
> (I_q) with $A = \alpha_q(M)$ and $B = \beta_q(M)$ is valid ?

We are now going to discuss these two programs. We will mainly be concerned with program $\mathcal{A}$ so we will just say some words about program $\mathcal{B}$.

4.1 PROGRAM $\mathcal{B}$

This program is in a great measure settled. As a first remark one can note that by proposition 3.9 we very easily get a positive answer to questions $1\mathcal{B}$ and $2\mathcal{B}$. More precisely, one has the following.

Proposition 4.1: *Let (M,g) be a compact Riemannian n-manifold. For any $1 \le q < n$, $\beta_q(M) = Vol_{(M,g)}^{-1/n}$ and $\beta_q(M) \in \mathcal{B}_q(M)$.*

Proof of proposition 4.1: Let $1 \le q < n$ be given. By taking $u = 1$ in (I_q) we get that $B \ge Vol_{(M,g)}^{-1/n}$. Hence, $\beta_q(M) \ge Vol_{(M,g)}^{-1/n}$. Independently, by proposition 3.9 there exists a positive constant $A = A(M, g, q)$ such that for any $u \in H_1^q(M)$,

$$\left(\int_M |u - \overline{u}|^p \, dv(g) \right)^{1/p} \le A \left(\int_M |\nabla u|^q \, dv(g) \right)^{1/q}$$

where $1/p = 1/q - 1/n$ and $\overline{u} = \frac{1}{Vol_{(M,g)}} \int_M u \, dv(g)$. As a consequence, for any $u \in H_1^q(M)$,

$$\left(\int_M |u|^p \, dv(g) \right)^{1/p} \le A \left(\int_M |\nabla u|^q \, dv(g) \right)^{1/q} + Vol_{(M,g)}^{1/p-1} \left| \int_M u \, dv(g) \right| ,$$

while, with Hölder,

$$\left| \int_M u \, dv(g) \right| \le Vol_{(M,g)}^{1-1/q} \left(\int_M |u|^q \, dv(g) \right)^{1/q}$$

Since $1/p - 1/q = -1/n$, we get that for any $u \in H_1^q(M)$,

$$\left(\int_M |u|^p \, dv(g) \right)^{1/p} \le A \left(\int_M |\nabla u|^q \, dv(g) \right)^{1/q} + Vol_{(M,g)}^{-1/n} \left(\int_M |u|^q \, dv(g) \right)^{1/q}$$

Since $\beta_q(M) \ge Vol_{(M,g)}^{-1/n}$, this ends the proof of the proposition.

Let us now briefly discuss question $3\mathcal{B}$. Answers to this question have been given by several authors. We mention Croke [Cr], Gallot [Gal3], [Gal4], and Ilias [I2]. As a consequence of their work, we have that when $B = \beta_q(M)$, the constant A of (I_q) mainly depends on a lower bound for the Ricci curvature, a lower bound for the volume, and an upper bound for the diameter. More precisely, one has the following.

Theorem 4.2: *Let (M,g) be a compact Riemannian n-manifold. Suppose that its Ricci curvature, volume, and diameter satisfy*

$$Rc_{(M,g)} \geq kg, \ Vol_{(M,g)} \geq v, \ and \ diam_{(M,g)} \leq d$$

where k, $v > 0$, and $d > 0$ are real numbers. Then, for any $1 \leq q < n$ and $1/p = 1/q - 1/n$, there exists a positive constant $A = A(n,q,k,v,d)$, depending only on n, q, k, v, and d, such that for any $u \in H_1^q(M)$,

$$\left(\int_M |u|^p dv(g)\right)^{1/p} \leq A\left(\left(\int_M |\nabla u|^q dv(g)\right)^{1/q} + Vol_{(M,g)}^{-1/n}\left(\int_M |u|^q dv(g)\right)^{1/q}\right)$$

Finally, about question $4\mathcal{B}$, we mention the following result of Ilias [I2]. (See also Bakry [Ba] and Fontenas [Fo] where the same result is obtained but in the more general context of abstract Markov generators).

Theorem 4.3: *Let (M,g) be a compact Riemannian n-manifold. Suppose that $Rc_{(M,g)} \geq (n-1)\delta g$ for some $\delta > 0$. Then for any $u \in H_1^2(M)$,*

$$\left(\int_M |u|^{2n/(n-2)} dv(g)\right)^{(n-2)/n}$$
$$\leq \frac{4}{n(n-2)\delta Vol_{(M,g)}^{2/n}}\int_M |\nabla u|^2 dv(g) + Vol_{(M,g)}^{-2/n}\int_M u^2 dv(g)$$

For more details on the subject we refer to Ilias [I2] (see in particular section 3.1), and Gallot [Gal4] (where isoperimetric inequalities based on integral norms of the Ricci curvature are obtained).

4.2 PROGRAM $\mathcal{A}$

Contrary to program $\mathcal{B}$, the difficulties appear here with the first question. Many authors have worked on this question. We mention Aubin [Au3], Federer - Fleming [FeF], Fleming - Rishel [FlR], Rosen [Ro], and Talenti [Ta]. As already mentioned, program $\mathcal{A}$ will be considered in the more general setting of complete manifolds. The first definitive and important result on the subject is then the following result published independently by Aubin [Au3] and Talenti [Ta]. We refer to [Au3], [Ta], or [Au6], for its proof.

Theorem 4.4: *Let $1 \le q < n$ and $1/p = 1/q - 1/n$.*

1) For any $u \in \mathcal{D}(\mathbf{R}^n)$,

$$\left(\int_{R^n} |u|^p dx\right)^{1/p} \le K(n,q) \left(\int_{R^n} |\nabla u|^q dx\right)^{1/q} \tag{6}$$

where

$$K(n,1) = \frac{1}{n}\left(\frac{n}{\omega_{n-1}}\right)^{1/n} ,$$

$$K(n,q) = \frac{1}{n}\left(\frac{n(q-1)}{(n-q)}\right)^{1-1/q} \left(\frac{\Gamma(n+1)}{\Gamma(n/q)\Gamma(n+1-n/q)\omega_{n-1}}\right)^{1/n} ,$$

and ω_{n-1} is the volume of the standard unit sphere of $\mathbf{R}^n$.

2) $K(n,q)$ is the best constant in (6) and if $q > 1$, the equality in (6) is attained by the functions

$$u(x) = \left(\frac{1}{\lambda + |x|^{q/(q-1)}}\right)^{n/q-1}$$

where λ is any positive real number.

Remarks: 1) When $q = 1$, (6) is the usual isoperimetric inequality [FeF], [FlR], [Fe]. A very nice proof of such an inequality is presented in Gromov [Gro]. (See also [Ch, section 6.2]). The extremum functions are here the characteristic functions of the balls of $\mathbf{R}^n$. When $q = 1$, (6) is sharp for an easy computation shows that

$$\left(\int_{R^n} |u_k|^{n/(n-1)} dx\right)^{(n-1)/n} \left(\int_{R^n} |\nabla u_k| dx\right)^{-1} = K(n,1)(1 + 0(1/k))$$

where the u_k's are defined by: $u_k(x) = 1$ when $0 \le |x| \le 1$, $u_k(x) = 1 + k(1 - |x|)$ when $1 \le |x| \le 1 + 1/k$, and $u_k(x) = 0$ when $|x| \ge 1 + 1/k$. Independently, note that $K(n,1)$ is the limiting value of $K(n,q)$ as $q \to 1$.

2) Several authors have conjectured that when $q = 1$, (6) holds for complete simply connected manifolds with nonpositive sectional curvature. It has been proved by Kleiner [Kl] that the conjecture is true if $n = 3$, and by Croke [Cr2] that the conjecture is true if $n = 4$. Recall here that by the work of Hoffman

and Spruck [HoS], one has that the $\mathbf{R}^n$-type H_1^1 Sobolev inequality is valid for any complete simply connected manifold with nonpositive sectional curvature.

3) Okikiolu [Ok], Glaser-Martin-Grosse-Thirring [GMGT], and Lieb [Lie] generalized theorem 4.4 when $q = 2$. One has that for any real number $0 \le b < 1$ and any $u \in \mathcal{D}(\mathbf{R}^n)$,

$$\left(\int_{R^n} |x|^{-bp} |u|^p dx \right)^{1/p} \le K_{n,p} \left(\int_{R^n} |\nabla u|^2 dx \right)^{1/2},$$

the equality being attained by the function

$$u(x) = \left(1 + |x|^{2t/r} \right)^{-r}$$

where $p = 2n/(2b + n - 2)$, $r = 2/(p - 2)$, $t = (n - 2)/2$,

$$K_{n,p} = \omega_{n-1}^{-(p-2)/2p} t^{-(p+2)/2p} M_p^{1/2},$$

and

$$M_p = \left((2r + 1)\Gamma(2r)/r\Gamma(r)^2 \right)^{1-2/p} (r/4)^{2/p} (r + 1)^{-1}$$

We refer the reader to [Lie] for more details.

With such a result, namely theorem 4.4, Aubin [Au3] was able to prove that $\alpha_q(M) = K(n, q)$ for compact manifolds and, more generally, for complete manifolds with bounded sectional curvature and positive injectivity radius. Surprisingly, contrary to $\beta_q(M)$, $\alpha_q(M)$ does not depend on the manifold. In Hebey [H4], we were able to prove that we still have $\alpha_q(M) = K(n, q)$ if the bound on the sectional curvature is replaced by a lower bound on the Ricci curvature. The result then becomes very sharp if one compares it with theorem 3.18 (see also corollary 3.19). In a certain sense, the optimal inequalities are valid as soon as the generic inequalities are valid. But before we go further, we prove the following folkloric result (a result everyone knows but which has no written proof).

Proposition 4.5: *Let (M, g) be a Riemannian n-manifold (not necessarily complete) and let $1 \le q < n$. Suppose that there exist $A, B \in \mathbf{R}$ such that for any $u \in \mathcal{D}(M)$,*

$$\left(\int_M |u|^p dv(g) \right)^{1/p} \le A \left(\int_M |\nabla u|^q dv(g) \right)^{1/q} + B \left(\int_M |u|^q dv(g) \right)^{1/q}$$

where $1/p = 1/q - 1/n$. Then $A \ge K(n, q)$, where $K(n, q)$ is as in theorem 4.4.

Proof of proposition 4.5: The proof is by contradiction. Suppose that there exist a Riemannian n-manifold (M, g) and real numbers $q \in [1, n)$, $A < K(n, q)$, and B, such that for any $u \in \mathcal{D}(M)$,

$$\left(\int_M |u|^p dv(g) \right)^{1/p} \le A \left(\int_M |\nabla u|^q dv(g) \right)^{1/q} + B \left(\int_M |u|^q dv(g) \right)^{1/q} \quad (7)$$

where $1/p = 1/q - 1/n$. Let $x \in M$. It is easy to see that for any $\epsilon > 0$ there exists a chart (Ω, ϕ) of M at x and there exists $\delta > 0$ such that $\phi(\Omega) = B_0^e(\delta) \subset \mathbf{R}^n$, and such that the components g_{ij} of g in this chart satisfy

$$(1 - \epsilon)\delta_{ij} \le g_{ij} \le (1 + \epsilon)\delta_{ij}$$

as bilinear forms. ($B_0^e(\delta)$ denotes the euclidean ball of center 0 and radius δ). Choosing ϵ small enough we then get by (7) that there exist $\delta_0 > 0$, $A' < K(n, q)$, and $B' \in \mathbf{R}$ such that for any $\delta \in (0, \delta_0)$ and any $u \in \mathcal{D}(B_0^e(\delta))$,

$$\left(\int_{R^n} |u|^p dx \right)^{1/p} \le A' \left(\int_{R^n} |\nabla u|^q dx \right)^{1/q} + B' \left(\int_{R^n} |u|^q dx \right)^{1/q}$$

But, by Hölder,

$$\left(\int_{B_0^e(\delta)} |u|^q dx \right)^{1/q} \le Vol_e \left(B_0^e(\delta) \right)^{1/n} \left(\int_{B_0^e(\delta)} |u|^p dx \right)^{1/p}$$

where e denotes the euclidean metric. Hence, choosing δ small enough we get that there exist $\delta > 0$ and $A'' < K(n, q)$ such that for any $u \in \mathcal{D}(B_0^e(\delta))$,

$$\left(\int_{R^n} |u|^p dx \right)^{1/p} \le A'' \left(\int_{R^n} |\nabla u|^q dx \right)^{1/q}$$

Let $u \in \mathcal{D}(\mathbf{R}^n)$. Set $u_\lambda(x) = u(\lambda x)$, $\lambda > 0$. For λ large enough, $u_\lambda \in \mathcal{D}(B_0^e(\delta))$. Hence,

$$\left(\int_{R^n} |u_\lambda|^p dx \right)^{1/p} \le A'' \left(\int_{R^n} |\nabla u_\lambda|^q dx \right)^{1/q}$$

But

$$\left(\int_{R^n} |u_\lambda|^p dx \right)^{1/p} = \lambda^{-n/p} \left(\int_{R^n} |u|^p dx \right)^{1/p}$$

while

$$\left(\int_{R^n} |\nabla u_\lambda|^q dx \right)^{1/q} = \lambda^{1-n/q} \left(\int_{R^n} |\nabla u|^q dx \right)^{1/q}$$

Since $1/p = 1/q - 1/n$, we get that for any $u \in \mathcal{D}(\mathbf{R}^n)$,

$$\left(\int_{R^n} |u|^p dx \right)^{1/p} \le A'' \left(\int_{R^n} |\nabla u|^q dx \right)^{1/q}$$

Since $A'' < K(n,q)$, such an inequality is in contradiction with theorem 4.4. This ends the proof of the proposition.

Remark: In order to prove proposition 4.5 one can also use truncated Bliss functions [Bl] brought to zero at the edge of a ball. Following Aubin [Au4], this argument is carried out explicitly in [HV3] for $q = 2$.

Let us now state Hebey's result [H4] we discussed few lines above. As already mentioned this result gives a sharp answer to question 1.A. It also furnishes an answer to question 3.A by showing that for $A \in A_q(M)$ close to $\alpha_q(M)$, the constant B of (I_q) mainly depends on a lower bound for the Ricci curvature and a lower bound for the injectivity radius. More precisely, we have the following.

Theorem 4.6: *Let (M,g) be a complete Riemannian n-manifold such that $Rc_{(M,g)} \geq kg$ for some $k \in \mathbf{R}$ and $inj_{(M,g)} \geq i$ for some $i > 0$. For any $\epsilon > 0$ and any $1 \leq q < n$ there exists a positive constant $B = B(\epsilon, n, q, k, i)$, depending only on ϵ, n, q, k, and i, such that for any $u \in H_1^q(M)$,*

$$\left(\int_M |u|^p dv(g) \right)^{q/p} \leq (K(n,q)^q + \epsilon) \int_M |\nabla u|^q dv(g) + B \int_M |u|^q dv(g)$$

where $1/p = 1/q - 1/n$ and $K(n,q)$ is as in theorem 4.4.

Note that since $q \geq 1$, $(x+y)^{1/q} \leq x^{1/q} + y^{1/q}$ for any $x, y \geq 0$. Combining theorem 4.6 with proposition 4.5 we then get the following.

Corollary 4.7: $\alpha_q(M) = K(n,q)$ *for any $1 \leq q < n$ and any complete Riemannian n-manifold with Ricci curvature bounded from below and positive injectivity radius.*

Of course, one also recovers the following result of Aubin [Au3].

Corollary 4.8: *Let (M,g) be a compact Riemannian n-manifold. For any $\epsilon > 0$ and any $1 \leq q < n$ there exists $B \in \mathbf{R}$ such that for any $u \in H_1^q(M)$,*

$$\left(\int_M |u|^p dv(g) \right)^{q/p} \leq (K(n,q)^q + \epsilon) \int_M |\nabla u|^q dv(g) + B \int_M |u|^q dv(g)$$

where $1/p = 1/q - 1/n$ and $K(n,q)$ is as in theorem 4.4.

Let us now prove theorem 4.6. The proof we present here is slightly different than the one of [H4]. The first step of the proof is the following result. (Think to compare it with lemma 3.17 and theorem 3.21).

Lemma 4.9: *Let (M,g) be a complete Riemannian n-manifold such that its Ricci curvature satisfies $Rc_{(M,g)} \geq kg$ for some $k \in \mathbf{R}$ and such that its injectivity radius satisfies $inj_{(M,g)} \geq i$ for some $i > 0$. For any $\epsilon > 0$ there exists a positive constant $\delta = \delta(\epsilon, n, k, i)$, depending only on ϵ, n, k, and i, such that for any $x \in M$, any $1 \leq q < n$, and any $u \in \mathcal{D}(B_x(\delta))$,*

$$\left(\int_M |u|^p dv(g) \right)^{q/p} \leq K(n,q)^q (1+\epsilon) \int_M |\nabla u|^q dv(g)$$

where $1/p = 1/q - 1/n$ and $K(n,q)$ is as in theorem 4.4.

Proof of lemma 4.9: By theorem 1.3 one easily sees that for any $\epsilon > 0$ there exists $\delta = \delta(\epsilon, n, k, i) > 0$ with the following property: for any $x \in M$ there exists a harmonic coordinate chart $\phi : B_x(\delta) \rightarrow \mathbf{R}^n$ such that the components g_{ij} of g in this chart satisfy

$$(1+\epsilon)^{-1} \delta_{ij} \leq g_{ij} \leq (1+\epsilon)\delta_{ij}$$

as bilinear forms. (Fix for instance $\alpha = 1/2$ in theorem 1.3). One then has that for any $x \in M$, any $1 \leq q < n$, and any $u \in \mathcal{D}(B_x(\delta))$,

$$\int_M |\nabla u|^q dv(g) \geq (1+\epsilon)^{-(n+q)/2} \int_{R^n} |\nabla(u \circ \phi^{-1})(x)|^q dx$$

and

$$\int_M |u|^p dv(g) \leq (1+\epsilon)^{n/2} \int_{R^n} |(u \circ \phi^{-1})(x)|^p dx$$

where $1/p = 1/q - 1/n$. Independently, by theorem 4.4,

$$\left(\int_{R^n} |(u \circ \phi^{-1})(x)|^p dx \right)^{q/p} \leq K(n,q)^q \int_{R^n} |\nabla(u \circ \phi^{-1})(x)|^q dx$$

As a consequence, we get that for any $\epsilon > 0$ there exists $\delta = \delta(\epsilon, n, k, i) > 0$ such that for any $x \in M$, any $1 \leq q < n$, and any $u \in \mathcal{D}(B_x(\delta))$,

$$\left(\int_M |u|^p dv(g) \right)^{q/p} \leq K(n,q)^q (1+\epsilon)^n \int_M |\nabla u|^q dv(g)$$

where $1/p = 1/q - 1/n$. This ends the proof of the lemma.

With such a result we are now in position to prove theorem 4.6.

Proof of theorem 4.6: Let $1 \leq q < n$ be given and let $p = nq/(n-q)$. By lemma 4.9 there exists $\delta = \delta(\epsilon, n, q, k, i) > 0$ such that for any $u \in \mathcal{D}(B_x(\delta))$,

$$\left(\int_M |u|^p dv(g) \right)^{q/p} \leq (K(n,q)^q + \epsilon/2) \int_M |\nabla u|^q dv(g)$$

Without loss of generality we can assume that $\delta < i$ for any $\epsilon > 0$. Independently, by lemma 1.6 we get that for any $\epsilon > 0$ there exists a sequence (x_j) of points of M such that:

(i) $M = \cup_j B_{x_j}(\delta/2)$ and $\forall j \neq j'$, $B_{x_j}(\delta/4) \cap B_{x_{j'}}(\delta/4) = \emptyset$

(ii) there exists $N = N(\epsilon, n, q, k, i)$ such that each point of M has a neighborhood which intersects at most N of the $B_{x_j}(\delta)$'s

where $\delta = \delta(\epsilon, n, q, k, i)$ is as above. Let $\eta_j = \alpha_j^{[q]+1}/\sum_m \alpha_m^{[q]+1}$ where the function $\alpha_j \in \mathcal{D}(B_{x_j}(\delta))$ is such that

$$0 \leq \alpha_j \leq 1, \ \alpha_j = 1 \text{ in } B_{x_j}(\delta/2), \text{ and } |\nabla \alpha_j| \leq 4/\delta$$

Clearly, (η_j) is a smooth partition of unity subordinate to the covering $(B_{x_j}(\delta))$, $\eta_j^{1/q} \in C^1(M)$ for any j, and there exists $H = H(\epsilon, n, q, k, i) > 0$ such that for any j, $|\nabla \eta_j^{1/q}| \leq H$. Fix $\epsilon > 0$ and let $u \in \mathcal{D}(M)$. We have that

$$\|u\|_p^q = \|u^q\|_{p/q} = \|\sum_j \eta_j u^q\|_{p/q} \leq \sum_j \|\eta_j u^q\|_{p/q} = \sum_j \|\eta_j^{1/q} u\|_p^q$$

where $\|u\|_p = \left(\int_M |u|^p dv(g)\right)^{1/p}$, while, for any j,

$$\|\eta_j^{1/q} u\|_p^q \leq (K(n,q)^q + \epsilon/2) \|\nabla(\eta_j^{1/q} u)\|_p^q$$

As a consequence we get that

$$\left(\int_M |u|^p dv(g)\right)^{q/p}$$

$$\leq (K(n,q)^q + \epsilon/2) \sum_j \int_M \left(\eta_j |\nabla u| + |u| |\nabla \eta_j^{1/q}|\right)^q dv(g)$$

$$\leq (K(n,q)^q + \epsilon/2) \int_M \sum_j \left(|\nabla u|^q \eta_j + \mu |\nabla u|^{q-1} |\nabla \eta_j^{1/q}| \eta_j^{(q-1)/q} |u| \right.$$

$$\left. + \nu |u|^q |\nabla \eta_j^{1/q}|^q\right) dv(g)$$

$$\leq (K(n,q)^q + \epsilon/2) \left(\|\nabla u\|_q^q + \mu N H \|\nabla u\|_q^{q-1} \|u\|_q + \nu N H^q \|u\|_q^q\right)$$

by Hölder's inequality and where μ and ν are such that for any $t \geq 0$ one has that $(1+t)^q \leq 1 + \mu t + \nu t^q$. (For instance, $\mu = q \max(1, 2^{q-2})$ and $\nu = \max(1, 2^{q-2})$). Now, let $\epsilon_0 > 0$ be such that

$$(K(n,q)^q + \epsilon/2)(1 + \epsilon_0) \leq K(n,q)^q + \epsilon$$

Since for any positive real numbers x, y, and λ, $qx^{q-1}y \le \lambda(q-1)x^q + \lambda^{1-q}y^q$, if we take $x = \|\nabla u\|_q$, $y = \|u\|_q$, and $\lambda = q\epsilon_0/\mu(q-1)NH$, we get that

$$\mu NH \|\nabla u\|_q^{q-1} \|u\|_q \le \epsilon_0 \|\nabla u\|_q^q + C \|u\|_q^q$$

where $C = \big(\mu NH/q\big)\big(q\epsilon_0/\mu(q-1)NH\big)^{1-q}$. Hence, for any $u \in \mathcal{D}(M)$,

$$\left(\int_M |u|^p dv(g) \right)^{q/p}$$

$$\le \big(K(n,q)^q + \epsilon/2\big)\big(1+\epsilon_0\big) \int_M |\nabla u|^q dv(g) + B \int_M |u|^q dv(g)$$

$$\le \big(K(n,q)^q + \epsilon\big) \int_M |\nabla u|^q dv(g) + B \int_M |u|^q dv(g)$$

where $B = \big(K(n,q)^q + \epsilon/2\big)\big(C + \nu NH^q\big)$. This ends the proof of the theorem.

We now concentrate on question 2.$\mathcal{A}$. We restrict ourselves to the case $q = 2$. The question then becomes on what conditions on a complete manifold (M, g) does one have that $\mathcal{A}_2(M)$ is closed? Aubin [Au3] simultaneously conjectured that $\mathcal{A}_2(M)$ is always closed for compact manifolds and he proved that $\mathcal{A}_2(M)$ is closed for complete manifolds with constant sectional curvature and positive injectivity radius. This result was then generalized by Hebey-Vaugon [HV1] to complete conformally flat manifolds with bounded sectional curvature and positive injectivity radius. We will present here two results obtained by Hebey-Vaugon in [HV2] and [HV3]. The main one, theorem 4.12, shows that $\mathcal{A}_2(M)$ is closed for complete manifolds with curvature bounded up to the order 1 and positive injectivity radius. As a consequence of this result, one gets that Aubin's conjecture is true (corollary 4.14). Roughly speaking, theorem 4.15 then shows that $\mathcal{A}_2(M)$ is closed for complete manifolds which are conformally flat at infinity. This generalizes the result mentioned above of [HV1]. Independently, note that from a PDE viewpoint it has till now been much more important to get optimal inequalities for $q = 2$ than for the other values of q. In particular, the knowledge of $\alpha_2(M)$ has played and still plays a very important role in the questions of existence of solutions to scalar curvature type equations, while the constant B of (I_2) when $A = \alpha_2(M)$ is connected to the existence of multiple solutions to these equations. We refer the reader to Hebey [H2] and the references therein for more details on these questions. (See also the historical notes at the end of the paragraph).

Before we state the two results mentioned above, we prove the following result and its main consequence for us (corollary 4.11). Given a Riemannian manifold (M, g), $Scal_{(M,g)}$ denotes its scalar curvature.

Proposition 4.10: *Let (M, g) be a Riemannian n-manifold of dimension $n \geq 4$ (not necessarily complete). Suppose that there exists $B \in \mathbf{R}$ such that for any $u \in \mathcal{D}(M)$,*

$$\left(\int_M |u|^{2n/(n-2)} dv(g) \right)^{(n-2)/n} \leq K(n,2)^2 \left(\int_M |\nabla u|^2 dv(g) + B \int_M u^2 dv(g) \right) \quad (8)$$

where $K(n,2)$ is as in theorem 4.4. Then, for any $x \in M$,

$$B \geq \frac{n-2}{4(n-1)} Scal_{(M,g)}(x)$$

Proof of proposition 4.10: We proceed as in Aubin [Au4]. Let $x \in M$ and let $r > 0$ be such that $r < inj_{(M,g)}(x)$ where $inj_{(M,g)}(x)$ is the injectivity radius at x. Then in geodesic normal coordinates

$$\frac{1}{\omega_{n-1}} \int_{S(r)} \sqrt{det(g_{ij})} \, ds = 1 - \frac{1}{6n} Scal_{(M,g)}(x) r^2 + O(r^4)$$

where $S(r) = \{ y \in M \text{ s.t. } d_g(x,y) = r \}$. For $\epsilon > 0$, we define

$$u_\epsilon = (\epsilon + r^2)^{1-n/2} - (\epsilon + \delta^2)^{1-n/2} \text{ if } r \leq \delta$$

$$u_\epsilon = 0 \text{ otherwise}$$

where $\delta \in (0, inj_{(M,g)}(x))$ is given and $r = d_g(x,.)$. Easy computations lead to

$$\int_M |\nabla u_\epsilon|^2 dv(g)$$

$$= \frac{(n-2)^2 \omega_{n-1}}{2} I_n^{n/2} \epsilon^{1-n/2} \left(1 - \frac{(n+2)}{6n(n-4)} Scal_{(M,g)}(x) \, \epsilon + o(\epsilon) \right) \text{ if } n > 4$$

$$= \frac{(n-2)^2 \omega_{n-1}}{2} \epsilon^{1-n/2} \left(I_n^{n/2} + \frac{1}{6n} Scal_{(M,g)}(x) \, \epsilon Log\epsilon + o(\epsilon Log\epsilon) \right) \text{ if } n = 4$$

$$\int_M u_\epsilon^2 dv(g)$$

$$= \frac{2(n-2)(n-1)\omega_{n-1}}{n(n-4)} I_n^{n/2} \epsilon^{2-n/2} + o(\epsilon^{2-n/2}) \text{ if } n > 4$$

$$= -\frac{\omega_{n-1}}{2} Log\epsilon + o(Log\epsilon) \text{ if } n = 4$$

$$\int_M u_\epsilon^{2n/(n-2)} dv(g)$$

$$\geq \frac{(n-2)\omega_{n-1}}{2n} I_n^{n/2} \epsilon^{-n/2} \left(1 - \frac{1}{6(n-2)} Scal_{(M,g)}(x) \, \epsilon + o(\epsilon) \right) \text{ if } n > 4$$

$$\geq \frac{(n-2)\omega_{n-1}}{2n} I_n^{n/2} \epsilon^{-n/2} \left(1 + o(\epsilon Log\epsilon) \right) \text{ if } n = 4$$

where $I_p^q = \int_0^{+\infty} (1+t)^{-p} t^q \, dt$. Independently, one easily checks that

$$\frac{\omega_n}{2^{n-1}\omega_{n-1}} = I_n^{n/2-1} = \frac{(n-2)}{n} I_n^{n/2}.$$

Hence,

$$\frac{(n-2)^2 \omega_{n-1}}{2} I_n^{n/2} = \frac{1}{K(n,2)^2} \left(\frac{(n-2)\omega_{n-1}}{2n} I_n^{n/2} \right)^{(n-2)/n}$$

and as a consequence of the developments made above we get that

$$\frac{\int_M |\nabla u_\epsilon|^2 dv(g) + B \int_M u_\epsilon^2 dv(g)}{\left(\int_M |u_\epsilon|^{2n/(n-2)} dv(g) \right)^{(n-2)/n}}$$

$$\leq \frac{1}{K(n,2)^2} \left(1 + \frac{\epsilon}{n(n-4)} \left(\frac{4(n-1)}{(n-2)} B - Scal_{(M,g)}(x) \right) + o(\epsilon) \right) \quad \text{if } n > 4$$

$$\leq \frac{1}{K(4,2)^2} \left(1 + \frac{1}{8} \left(Scal_{(M,g)}(x) - 6B \right) \epsilon Log\epsilon + o(\epsilon Log\epsilon) \right) \quad \text{if } n = 4$$

Since (8) implies that

$$\frac{\int_M |\nabla u_\epsilon|^2 dv(g) + B \int_M u_\epsilon^2 dv(g)}{\left(\int_M |u_\epsilon|^{2n/(n-2)} dv(g) \right)^{(n-2)/n}} \geq \frac{1}{K(n,2)^2},$$

we must have

$$\frac{4(n-1)}{(n-2)} B \geq Scal_{(M,g)}(x)$$

This ends the proof of the proposition.

Remark: When $q = 2$, $K(n,2) = \sqrt{\frac{4}{n(n-2)\omega_n^{2/n}}}$

As a consequence of proposition 4.10 we then get the following.

Corollary 4.11: *For any integer $n \geq 4$ there exist complete Riemannian n-manifolds with Ricci curvature bounded from below and positive injectivity radius for which there does not exist $B \in \mathbf{R}$ such that (8) is valid.*

Proof of corollary 4.11: Let (M,g) be a complete Riemannian n-manifold, $n \geq 4$, with Ricci curvature bounded from below. By proposition 4.10 the existence of $B \in \mathbf{R}$ such that (8) is valid implies that the Ricci curvature of (M,g) is bounded. (The existence of a lower bound for the Ricci curvature and of an upper bound for the scalar curvature obviously imply the existence of a bound for the Ricci curvature). Corollary 4.11 then comes from the existence of complete manifolds with Ricci curvature bounded from below and positive injectivity radius for which the Ricci curvature is not bounded. For the construction of such manifolds we refer to Anderson - Cheeger [AC].

Let us now state the main result which concerns question $2.A$. As one easily sees it also concerns question $3.A$. By corollary 4.11, stronger assumptions than those of theorem 4.6 are really necessary. The proof of theorem 4.12 is postponed to paragraph 4.3.

Theorem 4.12: *Let (M,g) be a complete Riemannian n-manifold such that $|Rm_{(M,g)}| \leq \Lambda_1$ and $|\nabla Rm_{(M,g)}| \leq \Lambda_2$ for some $\Lambda_1, \Lambda_2 \geq 0$, and such that $inj_{(M,g)} \geq i$ for some $i > 0$. There exists a positive constant $B = B(n, \Lambda_1, \Lambda_2, i)$, depending only on n, Λ_1, Λ_2, and i, such that for any $u \in H_1^2(M)$,*

$$\left(\int_M |u|^{2n/(n-2)} dv(g) \right)^{(n-2)/n} \leq K(n,2)^2 \int_M |\nabla u|^2 dv(g) + B \int_M u^2 dv(g)$$

where $K(n,2)$ is as in theorem 4.4.

As an easy consequence of this result one has the following.

Corollary 4.13: *Let (M,g) be a compact Riemannian n-manifold. For any Riemannian covering $(\tilde{M}, \tilde{g})$ of (M,g), this includes the trivial covering, there exists $B \in \mathbf{R}$ such that for any $u \in H_1^2(\tilde{M})$,*

$$\left(\int_{\tilde{M}} |u|^{2n/(n-2)} dv(\tilde{g}) \right)^{(n-2)/n} \leq K(n,2)^2 \int_{\tilde{M}} |\nabla u|^2 dv(\tilde{g}) + B \int_{\tilde{M}} u^2 dv(\tilde{g})$$

Corollary 4.14: *$\mathcal{A}_2(M)$ is a closed set for any compact Riemannian manifold.*

Remark: We asked in [HV3] if theorem 4.12 is still valid under the assumptions of theorem 4.6, namely $Rc_{(M,g)} \geq kg$ and $inj_{(M,g)} \geq i > 0$. By corollary 4.11 above, the answer to this question is no. Anyway, the problem is still open if one asks for theorem 4.12 to be valid under the assumptions $|Rc_{(M,g)}| \leq k$ and $inj_{(M,g)} \geq i > 0$. As mentioned in [HV3], a first more reasonnable goal would be to prove that theorem 4.12 is valid if one replaces the assumptions on $Rm_{(M,g)}$ by similar assumptions on $Rc_{(M,g)}$, namely if one asks for theorem 4.12 to be valid under the assumptions $|Rc_{(M,g)}| \leq \Lambda_1$, $|\nabla Rc_{(M,g)}| \leq \Lambda_2$, and $inj_{(M,g)} \geq i > 0$. One could for instance proceed as in [HV3], but with harmonic coordinates instead of geodesic normal coordinates. Recall, see theorem 1.3, that under the assumptions $|Rc_{(M,g)}| \leq \Lambda_1$, $|\nabla Rc_{(M,g)}| \leq \Lambda_2$, and $inj_{(M,g)} \geq i > 0$, one gets $C^{2,\alpha}$ bounds on the components of the metric tensor in harmonic coordinates.

Theorem 4.15: *Let (M,g) be a complete Riemannian n-manifold with bounded Ricci curvature and positive injectivity radius. Suppose that g is conformally flat outside a compact subset of M. Then there exists a constant $B \in \mathbf{R}$ such that for any $u \in H_1^2(M)$,*

$$\left(\int_M |u|^{2n/(n-2)} dv(g) \right)^{(n-2)/n} \leq K(n,2)^2 \int_M |\nabla u|^2 dv(g) + B \int_M u^2 dv(g)$$

where $K(n,2)$ is as in theorem 4.4.

Proof of theorem 4.15: Set $i = inj_{(M,g)}$ and let (x_j) be a sequence of points of M such that

(1) $M = \cup_j B_{x_j}(i/2)$ and $\forall j \neq j'$, $B_{x_j}(i/4) \cap B_{x_{j'}}(i/4) = \emptyset$

(2) there exists N integer such that each point of M has a neighborhood which intersects at most N of the $B_{x_j}(i)$'s.

The existence of such a sequence is given by lemma 1.6. Let $\eta_j = \alpha_j^2 / \sum_m \alpha_m^2$ where $\alpha_j \in \mathcal{D}(B_{x_j}(i))$ is such that

$$0 \leq \alpha_j \leq 1, \ \alpha_j = 1 \text{ in } B_{x_j}(i/2), \text{ and } |\nabla \alpha_j| \leq 4/i$$

Clearly (η_j) is a smooth partition of unity subordinate to the covering $\left(B_{x_j}(i) \right)$, $\eta_j^{1/2}$ is smooth for any j, and there exists $H > 0$ such that for any j, $|\nabla \eta_j^{1/2}| \leq H$.

Now, by hypothesis, there exists a compact subset K of M such that g is conformally flat on $M \backslash K$. Let j_0 be such that $B_{x_j}(i) \cap K = \emptyset$ for $j > j_0$. By theorem 4.12 one easily gets that for any $j \leq j_0$, there exists $B_j \in \mathbf{R}$ such that for any $u \in \mathcal{D}(M)$,

$$\left(\int_M |\eta_j^{1/2} u|^{2n/(n-2)} dv(g) \right)^{(n-2)/n}$$
$$\leq K(n,2)^2 \int_M |\nabla(\eta_j^{1/2} u)|^2 dv(g) + B_j \int_M \eta_j u^2 dv(g)$$

Independently, by proposition 3.25, for any $j > j_0$ and any $u \in \mathcal{D}(M)$,

$$\left(\int_M |\eta_j^{1/2} u|^{2n/(n-2)} dv(g) \right)^{(n-2)/n}$$
$$\leq K(n,2)^2 \int_M |\nabla(\eta_j^{1/2} u)|^2 dv(g) + S \int_M \eta_j u^2 dv(g)$$

where

$$S = \frac{1}{n(n-1)\omega_n^{2/n}} \, sup_M |Scal_{(M,g)}|$$

(S is finite since the Ricci curvature of (M,g) is assumed to be bounded). Set $B = max(B_1, \ldots, B_{j_0}, S)$. We then have that for any $u \in \mathcal{D}(M)$,

$$\left(\int_M |u|^{2n/(n-2)} dv(g) \right)^{(n-2)/n}$$

$$\leq K(n,2)^2 \sum_j \int_M |\nabla(\eta_j^{1/2} u)|^2 dv(g) + B \sum_j \int_M \eta_j u^2 dv(g)$$

$$= K(n,2)^2 \left(\sum_j \int_M \eta_j |\nabla u|^2 dv(g) + \sum_j \int_M u \nabla^\nu u \nabla_\nu \eta_j \, dv(g) \right.$$

$$\left. + \sum_j \int_M u^2 |\nabla \eta_j^{1/2}|^2 dv(g) \right) + B \sum_j \int_M \eta_j u^2 dv(g)$$

$$\leq K(n,2)^2 \int_M |\nabla u|^2 dv(g) + \left(B + K(n,2)^2 N H^2 \right) \int_M u^2 dv(g)$$

since $\sum_j \eta_j = 1$. This ends the proof of the theorem.

Remark: Let (M,g) be a compact Riemannian n-manifold. For $\alpha > 0$ real, set

$$\lambda(\alpha) = inf_{\{u \in C^\infty(M), \, u \not\equiv 0\}} \frac{\int_M |\nabla u|^2 dv(g) + \alpha \int_M u^2 dv(g)}{\left(\int_M |u|^{2n/(n-2)} dv(g) \right)^{(n-2)/n}}$$

By proposition 4.5, $\lambda(\alpha) \leq 1/K(n,2)^2$ for any $\alpha > 0$, while, by theorem 4.12, there exists A_0 depending on n, $Rm_{(M,g)}$, and $inj_{(M,g)}$ such that for $\alpha \geq A_0$, $\lambda(\alpha) = 1/K(n,2)^2$. (By proposition 4.10,

$$A_0 \geq ((n-2)/4(n-1)) \, max_{x \in M} Scal_{(M,g)}(x)$$

when $n \geq 4$. Independently, Gidas-Spruck [GiS] and Bidaut-Veron and Veron [BiV] proved that if the Ricci curvature of (M,g) satisfies $Rc_{(M,g)} \geq kg$ for some $k > 0$, then there exists a positive constant $a_0 = n(n-2)k/4(n-1)$ such that for $\alpha \in (0, a_0)$, $\lambda(\alpha) = \alpha Vol_{(M,g)}^{2/n}$. One then knows the exact value of $\lambda(\alpha)$ for small and large values of α. We will see in paragraph 4.4 (propositions 4.20 and 4.21) that for the standard unit sphere (S^n, h), $a_0 = A_0 = n(n-2)/4$.

Historical notes: Program $\mathcal{A}$ has its origin in 1960 with the work of Federer-Fleming [FeF] and Fleming-Rishel [FlR]. These authors proved that for every u belonging to a broad class of functions which vanish at infinity,

$$\left(\int_{R^n} |u|^{n/(n-1)} dx \right)^{(n-1)/n} \leq K(n,1) \int_{R^n} |\nabla u| dx$$

The sharp form of Sobolev inequality (6) when $n = 3$ and $q = 2$ was then discussed in 1971 by Rosen [Ro]. Finally, one had to wait 1976 and the work of Aubin [Au3] and Talenti [Ta] to obtain theorem 4.4 under the present form. As already mentioned, Aubin proved in addition that the value of the best constant α_q is still $K(n, q)$ when dealing with Sobolev inequalities on compact Riemannian manifolds (or complete Riemannian manifolds with bounded sectional curvature and positive injectivity radius). Independently, Aubin [Au4] proved that the best constant $\alpha_2 = K(n, 2)$ plays a fundamental role in the study of the Yamabe problem (and more generally in the study of scalar curvature type equations). This is what we are briefly going to discuss now.

The Yamabe problem can be stated as follows: given a compact Riemannian manifold (M, g) of dimension $n \geq 3$, prove that there exists a metric conformal to g of constant scalar curvature. From now on let (M, g) be a compact Riemannian n-manifold, $n \geq 3$. Any metric conformal to g can be written $g' = u^{4/(n-2)}g$ with $u \in C^{\infty}(M)$, $u > 0$. One then has that $Scal_{(M,g)}$ and $Scal_{(M,g')}$ satisfy the transformation law

$$\frac{4(n-1)}{n-2}\Delta_g u + Scal_{(M,g)}u = Scal_{(M,g')}u^{(n+2)/(n-2)}$$

Thus $g' = u^{4/(n-2)}g$ has constant scalar curvature λ if and only if u satisfies the Yamabe equation

$$\frac{4(n-1)}{n-2}\Delta_g u + Scal_{(M,g)}u = \lambda u^{(n+2)/(n-2)}, \ u > 0$$

In 1960, Yamabe [Y] attempted to solve this problem using techniques of calculus of variations and elliptic partial differential equations. He claimed that every compact Riemannian manifold has a conformal metric of constant scalar curvature. Unfortunately, his proof contained an error, discovered in 1968 by Trudinger [Tr]. Trudinger was able to repair the proof, but only with a rather restrictive assumption on the manifold. Finally, the problem was solved in two steps, by Aubin [Au4] in 1976 and Schoen [Sc1] in 1984. (See also the references [Sc2] and [ScY] which contain the proof of the positive mass theorem used in [Sc1]). This marked a milestone in the development of the theory of nonlinear partial differential equations. While semilinear equations of Yamabe type arise in many contexts and have long been studied by analysts, this was the first time that such an equation was completely solved. Set

$$I(u) = \frac{\frac{4(n-1)}{n-2}\int_M |\nabla u|^2 dv(g) + \int_M Scal_{(M,g)}u^2 dv(g)}{\left(\int_M |u|^{2n/(n-2)}dv(g)\right)^{(n-2)/n}}$$

As it can easily be seen, the positive critical points of I are smooth solutions of the Yamabe equation. Independently, by the Rellich-Kondrakov theorem one easily gets that for any $1 < q < 2n/(n-2)$ there exists $u_q \in C^\infty(M)$, $u_q > 0$, such that

$$\frac{4(n-1)}{n-2}\Delta_g u_q + Scal_{(M,g)}u_q = \lambda_q u_q^{q-1} \quad \text{and} \quad \int_M u_q^q dv(g) = 1$$

where

$$\lambda_q = inf_{\{u \neq 0 \text{ s.t. } \int_M u^q = 1\}}\left(\frac{4(n-1)}{n-2}\int_M |\nabla u|^2 dv(g) + \int_M Scal_{(M,g)}u^2 dv(g)\right)$$

This was first noticed by Yamabe. Yamabe's claim was then that a subsequence of the minimizing sequence (u_q) converges to a solution of the Yamabe equation as q tends to $2n/(n-2)$. Such a claim is false in general, unless, as pointed out by Aubin [Au4], one has that

$$inf_{\{u \neq 0\}} I(u) < \frac{4(n-1)}{n-2}K(n,2)^{-2}$$

This establishes the fundamental connection mentioned above that exists between the Yamabe problem and the value of the best constant $\alpha_2 = K(n,2)$. The result is sharp since one can prove that for any compact Riemannian n-manifold (M,g),

$$inf_{\{u \neq 0\}} I(u) \leq \frac{4(n-1)}{n-2}K(n,2)^{-2}$$

and that the equality is attained by the standard unit sphere (S^n, h). This result then shifts the focus of the proof from analysis to the problem of understanding the conformal invariant

$$\lambda(M) = inf_{\{u \neq 0\}} I(u)$$

The obvious approach to showing that

$$\lambda(M) < \lambda(S^n) = \frac{4(n-1)}{n-2}K(n,2)^{-2}$$

is to find a test function u such that $I(u) < \lambda(S^n)$. In 1976, Aubin [Au4] sought such a function compactly supported in a small neighborhood of a point $x \in M$. By carefully studying the local geometry of M near x he was able to construct such test functions, proving that if M has dimension $n \geq 6$ and is not conformally flat then $\lambda(M) < \lambda(S^n)$. The remaining cases were more difficult because the local geometry does not contain sufficient information to conclude that $\lambda(M) < \lambda(S^n)$. These cases thus require the construction of a global test

function. This was done by Schoen [Sc1] in 1984. In particular, he discovered
the unexpected relevance of the positive mass theorem of general relativity. His
theorem completes the solution of the Yamabe problem by showing that if M
has dimension 3, 4, or 5, or if M is conformally flat, then $\lambda(M) < \lambda(S^n)$ unless
(M, g) is conformal to (S^n, h). For more details on the subject we refer the
reader to Aubin [Au4] and Schoen [Sc1], to Schoen [Sc2] and Schoen-Yau [ScY]
for the proof of the positive mass theorem used in [Sc1], and more generally to
the surveys of Hebey [H2] and Lee-Parker [LeP].

4.3 PROOF OF THEOREM 4.12

The proof of theorem 4.12 is too long and too intricate to be entirely developed here. Our goal will just be to provide the reader with its main lines. As
one will see, the proof mixes PDE and geometric arguments. For more details
we refer the reader to Hebey-Vaugon [HV3] (see also [HV2]).

As a first remark, one can prove that the proof of theorem 4.12 reduces to
the proof of the following proposition. Roughly speaking, this can be proved
with similar packing arguments to those we used till now.

Proposition 4.16: *Let Λ_1 and Λ_2 be positive real numbers, and let g be a
smooth Riemannian metric on $\mathbf{R}^n$. Suppose that*

(i) $|Rm_{(\mathbf{R}^n, g)}| \leq \Lambda_1$ and $|\nabla Rm_{(\mathbf{R}^n, g)}| \leq \Lambda_2$ in $B_0^e(4)$

*(ii) the canonical coordinate system of $\mathbf{R}^n$, when restricted to $B_0^e(2)$, is a
normal geodesic coordinate system at 0 for g*

(iii) for any $x \in B_0^e(1)$, $min(\delta, inj_{(\mathbf{R}^n, g)}(x)) > 2$

*where $B_0^e(r)$ is the euclidean ball of $\mathbf{R}^n$ with center 0 and radius r, $inj_{(\mathbf{R}^n, g)}(x)$
is the injectivity radius of g at x, and $\delta = \delta(n, \Lambda_1, \Lambda_2)$ is as in lemma 1.4. Then
there exists a constant $B = B(n, \Lambda_1, \Lambda_2)$, depending only on n, Λ_1, and Λ_2, such
that for any $u \in \mathcal{D}(B_0^e(1))$,*

$$\left(\int_{B_0^e(1)} |u|^{2n/(n-2)} dv(g) \right)^{(n-2)/n}$$
$$\leq K(n, 2)^2 \int_{B_0^e(1)} |\nabla u|^2 dv(g) + B \int_{B_0^e(1)} u^2 dv(g)$$

The goal is then to prove proposition 4.16. Set $p = \frac{n+2}{n-2}$ and define, for any $\alpha > 0$ and any $u \in \mathcal{D}(B_0^e(1))$, $u \not\equiv 0$,

$$I_{g,\alpha}(u) = \frac{\int_{B_0^e(1)} |\nabla u|^2 dv(g) + \alpha \int_{B_0^e(1)} u^2 dv(g)}{\left(\int_{B_0^e(1)} |u|^{p+1} dv(g) \right)^{2/(p+1)}}$$

Proposition 4.16 can then be stated as follows: for any $n \geq 3$, $\Lambda_1 > 0$, and $\Lambda_2 > 0$, there exists $\alpha = \alpha(n, \Lambda_1, \Lambda_2)$ such that for any Riemannian metric g on $\mathbf{R}^n$ satisfying the points (i) to (iii) of proposition 4.16, and any $u \in \mathcal{D}(B_0^e(1))$, $u \not\equiv 0$, $I_{g,\alpha}(u) \geq K(n,2)^{-2}$. One just has to set $B = K(n,2)^2\alpha$. Now the proof of proposition 4.16 is by contradiction. Hence, we suppose that there exist $n \geq 3$, $\Lambda_1 > 0$, and $\Lambda_2 > 0$ such that the following holds: for any $\alpha > 0$ there exists a smooth Riemannian metric g_α on $\mathbf{R}^n$ such that

(iv) g_α satisfies the points (i) to (iii) of proposition 4.16

(v) $inf_{\{u \in \mathcal{D}(B_0^e(1)),\ u \not\equiv 0\}}\ I_{g_\alpha,\alpha}(u) < K(n,2)^{-2}$.

As a consequence of (iv) and lemma 1.4 one then has that there exists $K > 0$ independent of α such that for any $\alpha > 0$ and any $i, j, k = 1, \ldots, n$,

$$(1/4)\delta_{ij} \leq g_{ij}^\alpha \leq 4\delta_{ij} \text{ in } B_0^e(2) \text{ (as bilinear forms)}$$

$$\left| g_{ij}^\alpha \right| \leq K,\ \left| \partial_k g_{ij}^\alpha \right| \leq K \text{ in } B_0^e(2)$$

where the g_{ij}^α's are the components of g_α in the canonical chart of $\mathbf{R}^n$. Independently, as a consequence of (v) one easily gets by standard variational techniques that for any $\alpha > 0$ there exists $(\Phi_\alpha, \lambda_\alpha) \in C^2(\overline{B_0^e(1)}) \times (0, K(n,2)^{-2})$ such that

$$\Phi_\alpha > 0 \text{ in } B_0^e(1) \text{ and } \Phi_\alpha = 0 \text{ on } \partial B_0^e(1)$$

$$\Delta_{g_\alpha} \Phi_\alpha + \alpha \Phi_\alpha = \lambda_\alpha \Phi_\alpha^p \text{ in } B_0^e(1)$$

$$\int_{B_0^e(1)} \Phi_\alpha^{p+1} dv(g_\alpha) = 1$$

By standard concentration points arguments (see for instance Hebey [H1] for the approach used here), one can then prove that after passing to a subsequence:

$$\lim_{\alpha \to +\infty} \lambda_\alpha = K(n,2)^{-2} \text{ and } \lim_{\alpha \to +\infty} \int_{B_0^e(1)} \Phi_\alpha^2 dv(g_\alpha) = 0$$

(Φ_α) has one and only one concentration point $x \in \overline{B_0^e(1)}$

$$\lim_{\alpha \to +\infty} \Phi_\alpha = 0 \text{ in } C_{loc}^1(\overline{B_0^e(1)}\backslash\{x\})$$

$$\lim_{\alpha \to +\infty} \alpha \|\Phi_\alpha\|_{L^\infty(w)} = 0 \text{ for any } w \subset\subset \overline{B_0^e(1)}\backslash\{x\}$$

For convenience, set $u_\alpha = \left(\frac{\lambda_\alpha}{n(n-2)}\right)^{(n-2)/4}\Phi_\alpha$. Here again, (u_α) concentrates at x, and we have that

$$\Delta_{g_\alpha} u_\alpha + \alpha u_\alpha = n(n-2)u_\alpha^p \quad \text{in } B_0^e(1)$$

Let $x_\alpha \in B_0^e(1)$ be some point of $B_0^e(1)$ such that

$$u_\alpha(x_\alpha) = \|u_\alpha\|_{L^\infty\left(B_0^e(1)\right)}$$

and let $\mu_\alpha \in (0, +\infty)$ be such that

$$\|u_\alpha\|_{L^\infty\left(B_0^e(1)\right)} = \mu_\alpha^{-(n-2)/2}$$

According to what we have just said, we have that

$$\lim_{\alpha \to +\infty} x_\alpha = x \quad \text{and} \quad \lim_{\alpha \to +\infty} \mu_\alpha = 0$$

One can then prove, this is not straightforward, that after passing to a subsequence,

(vi) $\displaystyle \lim_{\alpha \to +\infty} \frac{d_e\left(x_\alpha, \partial B_0^e(1)\right)}{\mu_\alpha} = +\infty$

where d_e is the euclidean distance. Let $\Psi_\alpha = exp_{x_\alpha}$ be the exponential map of g_α at x_α. Set:

$$\omega_\alpha = \Psi_\alpha^{-1}\left(B_0^e(1)\right), \quad \tilde{g}_\alpha = (\Psi_\alpha)^* g_\alpha, \quad \text{and} \quad \tilde{u}_\alpha = u_\alpha \circ \Psi_\alpha$$

By (iv), $\tilde{g}_\alpha$ is defined in an open neighborhood of $\overline{B_0^e(2)}$ and, for any α, one has that $\overline{\omega}_\alpha \subset B_0^e(2)$. In addition, by lemma 1.4, there exists $\tilde{K} > 0$ independent of α such that for any α and any $i, j, k = 1, \ldots, n$,

(vii) $(1/4)\delta_{ij} \le \tilde{g}_{ij}^\alpha \le 4\delta_{ij}$ in $B_0^e(2)$ (as bilinear forms)

(viii) $\left|\tilde{g}_{ij}^\alpha\right| \le \tilde{K}, \quad \left|\partial_k \tilde{g}_{ij}^\alpha\right| \le \tilde{K}$ in $\overline{B_0^e(2)}$

where the $\tilde{g}_{ij}^\alpha$'s are the components of $\tilde{g}_\alpha$ in the canonical coordinate system of $\mathbf{R}^n$. Independently, one easily sees that

ω_α is starshaped at 0

$\forall i, j, k = 1, \ldots, n, \ \tilde{g}_{ij}^\alpha(0) = \delta_{ij}$ and $\partial_k \tilde{g}_{ij}^\alpha(0) = 0$

$\Delta_{\tilde{g}_\alpha} \tilde{u}_\alpha + \alpha \tilde{u}_\alpha = n(n-2)\tilde{u}_\alpha^p$ in ω_α

By (viii) and Ascoli we can now assume that $\displaystyle \lim_{\alpha \to +\infty} \tilde{g}_\alpha = \tilde{g}$ in $C^{0,1/2}\left(B_0^e(2)\right)$ for some $C^{0,1/2}$-Riemannian metric $\tilde{g}$ defined on $B_0^e(2)$. Set:

$$\tilde{v}_\alpha(y) = \mu_\alpha^{(n-2)/2}\tilde{u}_\alpha(\mu_\alpha y)$$

$$h_\alpha(y) = \tilde{g}_\alpha(\mu_\alpha y)$$

where $y \in \Omega_\alpha = \{\mu_\alpha^{-1}z, \ z \in \omega_\alpha\}$. According to what we have said till now we easily get that

$$0 \le \tilde{v}_\alpha \le 1, \ \tilde{v}_\alpha(0) = 1, \ \text{and} \ \Delta_{h_\alpha}\tilde{v}_\alpha + (\alpha\mu_\alpha^2)\tilde{v}_\alpha = n(n-2)\tilde{v}_\alpha^p \ \text{in} \ \Omega_\alpha$$

$$\lim_{\alpha \to +\infty} d_e(0, \partial\Omega_\alpha) = +\infty$$

$$\forall \Theta \subset\subset \mathbf{R}^n, \ \lim_{\alpha \to +\infty} h_\alpha = e \ \text{in} \ C^1(\Theta)$$

where e is the euclidean metric of $\mathbf{R}^n$. One then gets that the $\tilde{v}_\alpha$'s are equicontinuous on every compact subset of $\mathbf{R}^n$. Hence, by Ascoli, there exists $\tilde{v} \in C^0(\mathbf{R}^n)$ such that for any $\Theta \subset\subset \mathbf{R}^n$, a subsequence of $(\tilde{v}_\alpha)$ converges to $\tilde{v}$ in $L^\infty(\Theta)$. This allows one to prove that $\lim_{\alpha \to +\infty} \alpha\mu_\alpha^2 = 0$, and as a consequence we get that $\tilde{v} \in C^\infty(\mathbf{R}^n)$ and that

$$\Delta_e\tilde{v} = n(n-2)\tilde{v}^p \ \text{in} \ \mathbf{R}^n$$

By Caffarelli-Gidas-Spruck [CaGS], see also Obata [Ob], we then get that

$$\tilde{v}(y) = \left(1 + |y|^2\right)^{-(n-2)/2}$$

where $|y| = d_e(0, y)$, d_e the euclidean distance. The fundamental estimate which will allow us to obtain the desired contradiction is then the following. Note that as a consequence of this estimate, one has that $\lim_{\alpha \to +\infty} \tilde{v}_\alpha = \tilde{v}$ in $L^\infty(\mathbf{R}^n)$.

Lemma 4.17: *There exists $C > 0$, independent of α, such that for any $\alpha \gg 1$ and any $y \in \Omega_\alpha$, $\tilde{v}_\alpha(y) \le C\tilde{v}(y)$.*

With such an estimate we are now in position to prove proposition 4.16. As already mentioned, this will end the proof of theorem 4.12. (Recall that we are looking for a contradiction). By the standard Pohozaev identity, one has that for any α,

$$\int_{\Omega_\alpha} (y^k\partial_k\tilde{v}_\alpha)\Delta_e\tilde{v}_\alpha dx + \frac{n-2}{2}\int_{\Omega_\alpha} \tilde{v}_\alpha\Delta_e\tilde{v}_\alpha dx = -\frac{1}{2}\int_{\partial\Omega_\alpha} (y, \nu)_e(\partial_\nu\tilde{v}_\alpha)^2 ds$$

where $(.,.)_e$ is the euclidean scalar product and ν is the unit outer normal to $\partial\Omega_\alpha$. Since Ω_α is starshaped at 0 one then gets that for any α.

$$\int_{\Omega_\alpha} (y^k\partial_k\tilde{v}_\alpha)\Delta_e\tilde{v}_\alpha dx + \frac{n-2}{2}\int_{\Omega_\alpha} \tilde{v}_\alpha\Delta_e\tilde{v}_\alpha dx \le 0 \tag{ix}$$

Let us now write that

$$\frac{1}{4}\Delta_e \tilde{v}_\alpha = \Delta_{h_\alpha} \tilde{v}_\alpha + \left(h_\alpha^{ij} - \frac{1}{4}\delta_{ij}\right)\partial_{ij}\tilde{v}_\alpha - h_\alpha^{ij}\Gamma(h_\alpha)_{ij}^k \partial_k \tilde{v}_\alpha \qquad (x)$$

where (h_α^{ij}) is the inverse matrix of (h_{ij}^α) and the $\Gamma(h_\alpha)_{ij}^k$'s are the Christoffel symbols of h_α. Combining (ix) and (x), the equations satisfied by the $\tilde{v}_\alpha$'s, and estimates on the h_α's, one then gets, after laborious developments, that there exist $C_1 > 0$ and $C_2 > 0$, independent of α, such that for any α,

$$\alpha \int_{\Omega_\alpha} \tilde{v}_\alpha^2 dx \le C_1 \int_{\Omega_\alpha} \tilde{v}_\alpha^2 dx + C_2 \int_{\Omega_\alpha} |y|^2 \tilde{v}_\alpha^{p+1} dx \qquad (xi)$$

Now, note that since $\lim_{\alpha \to +\infty} \tilde{v}_\alpha = \tilde{v}$ in $L^\infty(\mathbf{R}^n)$, we have that for $\alpha \gg 1$

$$\int_{\Omega_\alpha} \tilde{v}_\alpha^2 dx \ge \int_{B_0^c(1)} \tilde{v}^2 dx > 0 \ ,$$

while by lemma 4.17 we get the existence of positive constants C_3 and C_4 independent of α such that for $\alpha \gg 1$

$$\int_{\Omega_\alpha} |y|^2 \tilde{v}_\alpha^{p+1} dx \le C_3 \int_{\mathbf{R}^n} |y|^2 \tilde{v}^{p+1} dx \le C_4$$

Combining (xi) with these two inequalities one then easily obtains the contradiction we were looking for. As already mentioned, this ends the proof of theorem 4.12.

Remark: Very recently, this approach has been used by YanYan Li and Meijun Zhu [LiZ] to prove that the sharp Sobolev trace inequality holds on compact Riemannian manifolds with boundary. More precisely, Li and Zhu prove that for any compact Riemannian n-manifold (M, g) with smooth boundary ∂M, there exists some positive constant B such that for any $u \in H_1^2(M)$,

$$\left(\int_{\partial M} |u|^{2(n-1)/(n-2)} ds\right)^{(n-2)/(n-1)} \le S \int_M |\nabla u|^2 dv(g) + B \int_{\partial M} u^2 ds$$

where ds denotes the induced volume element on ∂M, and $S = \dfrac{2}{(n-2)\omega_{n-1}^{1/(n-1)}}$ is the best constant for such an inequality.

4.4 PROGRAM $\mathcal{A}$ - VARIOUS RESULTS AND OPEN QUESTIONS

After what we have said in paragraph 4.2, several natural questions arise. For instance:

Question 1: for compact manifolds, is $\mathcal{A}_q(M)$ a closed set for any $1 \le q < n$?

Question 2: what happens to theorem 4.12 for $1 \le q < n$, $q \ne 2$?

Question 3: is theorem 4.12 valid under the assumptions $|Rc_{(M,g)}| \le k$ and $inj_{(M,g)} \ge i > 0$?

Partial answers to questions 1 and 2 are given by the two following results of Aubin [Au3]. Note that (as suggested by theorem 4.19) there is perhaps a significant difference between the optimal inequality

$$\left(\int_M |u|^p dv(g) \right)^{1/p} \le K(n,q) \left(\int_M |\nabla u|^q dv(g) \right)^{1/q} + B \left(\int_M |u|^q dv(g) \right)^{1/q}$$

and the stronger optimal inequality

$$\left(\int_M |u|^p dv(g) \right)^{q/p} \le K(n,q)^q \int_M |\nabla u|^q dv(g) + B \int_M |u|^q dv(g)$$

Theorem 4.18: *Let (M,g) be a complete Riemannian n-manifold with constant sectional curvature and positive injectivity radius. For any $1 \le q < n$ and $1/p = 1/q - 1/n$ there exists $B \in \mathbf{R}$ such that for any $u \in H_1^q(M)$,*

$$\left(\int_M |u|^p dv(g) \right)^{1/p} \le K(n,q) \left(\int_M |\nabla u|^q dv(g) \right)^{1/q} + B \left(\int_M |u|^q dv(g) \right)^{1/q}$$

where $K(n,q)$ is as in theorem 4.4.

Theorem 4.19: *Let (S^n, h) be the standard unit sphere of $\mathbf{R}^{n+1}$. For any $1 \le q < n$ and $1/p = 1/q - 1/n$ there exists $B \in \mathbf{R}$ such that for any $u \in H_1^q(S^n)$,*

$$\left(\int_{S^n} |u|^p dv(h) \right)^{q/p} \le K(n,q)^q \int_{S^n} |\nabla u|^q dv(h) + B \int_{S^n} |u|^q dv(h)$$

if $1 \le q \le 2$, and

$$\left(\int_{S^n} |u|^p dv(h) \right)^{q/p(q-1)}$$
$$\le K(n,q)^{q/(q-1)} \left(\int_{S^n} |\nabla u|^q dv(h) \right)^{1/(q-1)} + B \left(\int_{S^n} |u|^q dv(h) \right)^{1/(q-1)}$$

if $2 \le q < n$, where $K(n,q)$ is as in theorem 4.4.

Let us now discuss question 4A. Here few things are known and results such as theorem 4.3 are missing. Let $(\mathbf{R}^n, e)$ be the n-euclidean space, (S^n, h)

be the standard unit sphere of $\mathbf{R}^{n+1}$, (H^n, h') be the n-dimensional simply connected hyperbolic space, and (P^n, p) be the n-dimensional projective space with its canonical metric induced from h. Consider the following inequality: $\forall u \in H_1^2(M)$,

$$\left(\int_M |u|^{2n/(n-2)} dv(g) \right)^{(n-2)/n} \leq K(n,2)^2 \int_M |\nabla u|^2 dv(g) + B \int_M u^2 dv(g) \quad (9)$$

By Hebey-Vaugon [HV1] and [HV3], one has the following. The second part of proposition 4.20 comes from proposition 4.10.

Proposition 4.20: (9) *is valid with*

1) $B = 0$ *for* $(\mathbf{R}^n, e)$, $B = -\frac{1}{\omega_n^{2/n}}$ *for* (H^n, h'), *and* $B = \frac{1}{\omega_n^{2/n}}$ *for* (S^n, h), $n \geq 3$

2) $B = \frac{(n+2)}{(n-2)\omega_n^{2/n}}$ *for* (P^n, p), *and* $B = \frac{1+(n-2)^2}{n(n-2)\omega_n^{2/n}}$ *for* $(S^1 \times S^{n-1}, h \times h)$, $n \geq 3$

3) $B = \frac{m-n}{(m+n)\omega_{m+n}^{2/(m+n)}}$ *for* $(S^m \times H^n, h \times h')$, $m \geq 2, n \geq 2$

4) $B = \frac{m-n+2}{(m+n-2)\omega_{m+n}^{2/(m+n)}}$ *for* $(P^m \times H^n, p \times h')$, $m \geq 2, n \geq 2$

5) $B = \frac{n-1}{(n+1)\omega_{n+1}^{2/(n+1)}}$ *for* $(S^n \times \mathbf{R}, h \times e)$, $B = \frac{n+1}{(n-1)\omega_{n+1}^{2/(n+1)}}$ *for* $(P^n \times \mathbf{R}, p \times e)$, *and* $B = -\frac{(n-1)}{(n+1)\omega_{n+1}^{2/(n+1)}}$ *for* $(H^n \times \mathbf{R}, h' \times e)$, $n \geq 2$

Furthermore, at least when the dimension of the manifold is greater than or equal to 4, these inequalities are totally optimal (in the sense that B can not be lowered) for $(\mathbf{R}^n, e)$, (H^n, h'), (S^n, h), $(S^m \times H^n, h \times h')$, $(S^n \times \mathbf{R}, h \times e)$, *and* $(H^n \times \mathbf{R}, h' \times e)$.

Proof of proposition 4.20: With the exception of $S^1 \times S^{n-1}$, P^n, $P^m \times H^n$, and $P^n \times \mathbf{R}$, all the manifolds listed in proposition 4.20 are conformally flat and simply connected. For such manifolds, one easily obtains from corollary 3.26 (and proposition 4.10) what is stated in proposition 4.20. On the other hand, since they are not anymore simply connected, the situation is a little bit more delicate for $S^1 \times S^{n-1}$, P^n, $P^m \times H^n$, and $P^n \times \mathbf{R}$. We present the proof for P^n and refer to [HV1] and [HV3] for $S^1 \times S^{n-1}$, $P^m \times H^n$, and $P^n \times \mathbf{R}$. By [HV1], there exists a covering $(\Omega_i)_{i=1,...,n+1}$ of P^n and a smooth partition of unity $(\eta_i)_{i=1,...,n+1}$ such that

(i) for any i, Ω_i is simply connected

(ii) for any i, η_i and $\sqrt{\eta_i} \in \overset{\circ}{H}_1^2(\Omega_i)$

(iii) $\displaystyle\sum_{i=1}^{n+1} |\nabla\sqrt{\eta_i}|^2 = n$

Let $u \in C^\infty(P^n)$. As in the proof of theorem 4.6, one has that

$$\left(\int_{P^n} |u|^{2n/(n-2)} dv(p)\right)^{(n-2)/n} \leq \sum_{i=1}^{n+1}\left(\int_{P^n} |\sqrt{\eta_i}\,u|^{2n/(n-2)} dv(p)\right)^{(n-2)/n}$$

Since Ω_i is simply connected and since the scalar curvature of P^n equals $n(n-1)$, one gets by corollary 3.26 that for any i,

$$\left(\int_{P^n} |\sqrt{\eta_i}\,u|^{2n/(n-2)} dv(p)\right)^{(n-2)/n}$$

$$\leq K(n,2)^2 \int_{P^n} |\nabla(\sqrt{\eta_i}\,u)|^2 dv(p) + \omega_n^{-2/n}\int_{P^n} \eta_i u^2 dv(p)$$

Hence, since $\displaystyle\sum_{i=1}^{n+1}\eta_i = 1$ and $\displaystyle\sum_{i=1}^{n+1}|\nabla\sqrt{\eta_i}|^2 = n$,

$$\left(\int_{P^n} |u|^{2n/(n-2)} dv(p)\right)^{(n-2)/n}$$

$$\leq \sum_{i=1}^{n+1}\left(\int_{P^n} |\sqrt{\eta_i}\,u|^{2n/(n-2)} dv(p)\right)^{(n-2)/n}$$

$$\leq K(n,2)^2 \sum_{i=1}^{n+1}\int_{P^n} |\nabla(\sqrt{\eta_i}\,u)|^2 dv(p) + \omega_n^{-2/n}\sum_{i=1}^{n+1}\int_{P^n} \eta_i u^2 dv(p)$$

$$= K(n,2)^2 \sum_{i=1}^{n+1}\int_{P^n}\left(\eta_i|\nabla u|^2 + |\nabla\sqrt{\eta_i}|^2 u^2 + u\nabla^\nu\eta_i\nabla_\nu u\right)dv(p)$$

$$+ \omega_n^{-2/n}\int_{P^n} u^2 dv(p)$$

$$= K(n,2)^2 \int_{P^n} |\nabla u|^2 dv(p) + nK(n,2)^2\int_{P^n} u^2 dv(p)$$

$$+ \omega_n^{-2/n}\int_{P^n} u^2 dv(p)$$

$$= K(n,2)^2 \int_{P^n} |\nabla u|^2 dv(p) + \frac{(n+2)}{(n-2)\omega_n^{2/n}}\int_{P^n} u^2 dv(p)$$

This ends the proof of the proposition.

Here again, natural questions arise. For instance:

Question 4: can one find large classes of manifolds, in the spirit of theorem 4.3, for which the constant B of (9) is computable ?

Question 5: what is the value of $B_{opt} = inf\{B \text{ s.t. (9) is valid}\}$ for the projective space ?

By proposition 4.20 and what we have said in paragraph 4.1, $B_{opt}(P^n)$ is greater than or equal to $(2/\omega_n)^{2/n}$ and less than or equal to $(n+2)/((n-2)\omega_n^{2/n})$. In addition, one has that $\lim_{n\to+\infty}(n+2)/(2^{2/n}(n-2)) = 1$.

Let us now discuss the final question of programs $\mathcal{A}$ and $\mathcal{B}$. Here again we will mainly be concerned with the case $q = 2$. The question we ask is the following: for which compact Riemannian n-manifolds without boundary (M, g) does one have that for any $u \in H_1^2(M)$,

$$\left(\int_M |u|^{2n/(n-2)} dv(g)\right)^{(n-2)/n}$$
$$\leq K(n,2)^2 \int_M |\nabla u|^2 dv(g) + Vol_{(M,g)}^{-2/n} \int_M u^2 dv(g) \qquad (10)$$

Such an inequality is totally optimal in the sense that the two constants $K(n,2)^2$ and $Vol_{(M,g)}^{-2/n}$ can not be lowered. The first result concerning this question is the following result of Aubin [Au4]. (By (10) is valid we mean that (10) holds for any $u \in H_1^2(M)$).

Proposition 4.21: (10) *is valid for the standard unit sphere* (S^n, h).

The guess here (call it a conjecture if you want) is that (S^n, h) is the only compact Riemannian manifold for which (10) is valid. More precisely, one should have that if (10) is valid for a compact Riemannian n-manifold (M, g), then, up to a constant scale factor, (M, g) and (S^n, h) are isometric. A beginning of answer has been obtained in Hebey-Vaugon [HV1] where the following is proved. We refer the reader to [HV1] for a more precise statement.

Proposition 4.22: *Let* (M, g) *be a compact Riemannian n-manifold, and let* $\lambda_1(g)$ *be the first nonzero eigenvalue of* Δ_g. *Assume that the scalar curvature of* (M, g) *is constant.*

1) *If (10) is valid,* $\lambda_1(g) \geq n\left(\frac{\omega_n}{Vol_{(M,g)}}\right)^{2/n}$

2) *Conversely, if* $\lambda_1(g) > n\left(\frac{\omega_n}{Vol_{(M,g)}}\right)^{2/n}$ *there exists an infinite number of non-homothetic functions* $u \in H_1^2(M)$ *such that*

$$\left(\int_M |u|^{2n/(n-2)} dv(g)\right)^{(n-2)/n} \leq K(n,2)^2 \int_M |\nabla u|^2 dv(g) + Vol_{(M,g)}^{-2/n} \int_M u^2 dv(g)$$

As a consequence of this result, (10) is satisfied by an infinite number of non-homothetic functions on the projective space (P^n, p). (Independently, it is possible to prove that there exist quotients of (S^n, h) for which (10) is not valid. Think for instance to $S^3 \subset \mathbf{R}^4 = \mathbf{C}^2$, $u_k(z, z') = |z|^2$, $G_k = \{\sigma_k^j\}$, $\sigma_k(z, z') = e^{2i\pi/k}(z, z')$, $P_k = S^n/G_k$, and let $k \to +\infty$).

Independently, we get as an easy consequence of proposition 4.10 the following result. If (M, g) is a compact Riemannian n-manifold, the Yamabe energy of (M, g), denoted by $Yam(M, g)$, is defined by

$$Yam(M, g) = \frac{1}{Vol_{(M,g)}^{(n-2)/n}} \int_M Scal_{(M,g)} dv(g)$$

Proposition 4.23: *Let (M, g) be a compact Riemannian n-manifold, $n \geq 4$. If (10) is valid, then $Yam(M, g) \leq Yam(S^n, h)$.*

Combining proposition 4.23 and the well-known fact that for any g in the conformal class of h one has that $Yam(S^n, g) \geq Yam(S^n, h)$, with equality if and only if g has constant sectional curvature (see for instance Lee-Parker [LeP, theorem 3.2]), we then get that the conjecture mentioned above is valid in the conformal class of the standard metric h of S^n, $n \geq 4$.

Corollary 4.24: *Let g be a Riemannian metric on S^n, $n \geq 4$, conformal to the standard metric h. If (10) is valid for (S^n, g) then, up to a constant scale factor, g and h are isometric.*

Independently, combining proposition 4.22 and proposition 4.23, we get the following.

Corollary 4.25: *Let (M, g) be a compact Riemannian n-manifold, $n \geq 4$, and let $\lambda_1(g)$ be the first non-zero eigenvalue of Δ_g. Assume that the scalar curvature of (M, g) is constant. If (10) is valid, then $\lambda_1(g) \geq \frac{1}{(n-1)} Scal_{(M,g)}$.*

Remark: According to Aubin [Au4] such an inequality is also satisfied by Yamabe metrics. (A Yamabe metric is a metric which realizes the infimum of the Yamabe energy in its conformal class. Thanks to the resolution of the Yamabe problem by Aubin and Schoen, every conformal class contains a Yamabe metric). One also has that for any Yamabe metric g on a compact manifold M, $Yam(M, g) \leq Yam(S^n, h)$.

Let us now mention that as an easy consequence of Bakry-Ledoux [BaL, theorem 4], one has the following.

Proposition 4.26: Let (M, g) be a compact Riemannian n-manifold. If (10) is valid, then $diam_{(M,g)} Vol_{(M,g)}^{-1/n} \leq \pi \omega_n^{-1/n}$.

Proof of proposition 4.26: The proof is by contradiction. Suppose that (10) is valid and that $diam_{(M,g)} Vol_{(M,g)}^{-1/n} > \pi \omega_n^{-1/n}$. Since (10) and $diam_{(M,g)} Vol_{(M,g)}^{-1/n}$ are obviously scale invariant, after rescaling the metric we can assume that $diam_{(M,g)} = \pi$. As a consequence, $Vol_{(M,g)}^{-1/n} > \omega_n^{-1/n}$ and we get that for any $u \in H_1^2(M)$,

$$\left(\int_M |u|^{2n/(n-2)} dv(g) \right)^{(n-2)/n}$$
$$\leq \frac{4}{n(n-2) Vol_{(M,g)}^{2/n}} \int_M |\nabla u|^2 dv(g) + Vol_{(M,g)}^{-2/n} \int_M u^2 dv(g)$$

We can now apply theorem 4 of Bakry-Ledoux [BaL] and we get that there exist some non-constant function $u_0 \in H_1^2(M)$ such that

$$\left(\int_M |u_0|^{2n/(n-2)} dv(g) \right)^{(n-2)/n}$$
$$= \frac{4}{n(n-2) Vol_{(M,g)}^{2/n}} \int_M |\nabla u_0|^2 dv(g) + Vol_{(M,g)}^{-2/n} \int_M u_0^2 dv(g)$$

Independently, since (10) is valid, one also has that

$$\left(\int_M |u_0|^{2n/(n-2)} dv(g) \right)^{(n-2)/n}$$
$$\leq \frac{4}{n(n-2) \omega_n^{2/n}} \int_M |\nabla u_0|^2 dv(g) + Vol_{(M,g)}^{-2/n} \int_M u_0^2 dv(g)$$

The fact that $\int_M |\nabla u_0|^2 dv(g) \neq 0$ then implies that $Vol_{(M,g)}^{-2/n} \leq \omega_n^{-2/n}$, which is the contradiction we were looking for. This ends the proof of the proposition.

Remark: Let (M, g) be a compact Riemannian n-manifold. For $k \geq 1$ an integer, set $M_k = M \times \ldots \times M$ (k times), and let $g_k = g \times \ldots \times g$ be the product metric on M_k. It is easy to see that $diam_{(M_k, g_k)} Vol_{(M_k, g_k)}^{-1/nk} = \sqrt{k} diam_{(M,g)} Vol_{(M,g)}^{-1/n}$. Hence, by proposition 4.26, for any compact Riemannian n-manifold (M, g) there exists some integer $k_0 \geq 1$ such that for any $k \geq k_0$, (10) is not valid for (M_k, g_k). Since $Yam(M_k, g_k) = k Yam(M, g)$, such a result can also be deduced from proposition 4.23 when $Yam(M, g) > 0$.

As a straightforward consequence of corollary 3.12 one also has the following.

Proposition 4.27: *For any* $\Lambda > 0$ *there are only finitely many diffeomorphism types of compact Riemannian n-manifolds* (M, g) *such that simultaneously* $|K_{(M,g)}|Vol_{(M,g)}^{2/n} \leq \Lambda$ *and (10) is valid.*

Finally, we mention the following result of Beckner [Be]. It concerns the extension of (10) on (S^n, h) to powers $2 \leq p \leq 2n/(n-2)$. We refer to [Be] for the proof of this result. (See also Bakry [Ba] and Fontenas [Fo] where the result is obtained in the more general context of abstract Markov generators).

Theorem 4.28: *Let* (S^n, h) *be the standard unit sphere of* $\mathbf{R}^{n+1}$. *For any* $2 \leq p \leq 2n/(n-2)$ *and any* $u \in H_1^2(S^n)$,

$$\left(\int_{S^n} |u|^p dv(h) \right)^{2/p} \leq \frac{(p-2)}{n\omega_n^{1-2/p}} \int_{S^n} |\nabla u|^2 dv(h) + \frac{1}{\omega_n^{1-2/p}} \int_{S^n} u^2 dv(h)$$

Remark: The extremum functions for (10) on (S^n, h) are well-known. One can find their expression in Aubin [Au6, theorem 6.12]. We refer also to Bakry-Ledoux [BaL] and Beckner [Be].

4.5 IMPROVEMENT OF THE BEST CONSTANTS

We discuss in this paragraph a direction of research first developed by Aubin [Au5]. The general idea is to show that orthogonality conditions allow one to lower the best constant $K(n, q)$. The main difference between the results we present here and those of Aubin comes from theorems 4.6 and 4.12. Concerning theorem 4.29, the assumption made by Aubin that the sectional curvature of the manifold is bounded is replaced by a lower bound on the Ricci curvature. Concerning theorem 4.30, the assumption made by Aubin that the manifold has constant sectional curvature is replaced by bounds on the Riemann curvature. For examples of applications of the following type results to the study of partial differential equations, we refer to Aubin [Au5], Chang-Yang [ChYa], and Hebey [H3]. In the following, $K(n, q)$ is as in theorem 4.4.

Theorem 4.29: *Let* (M, g) *be a complete Riemannian n-manifold with Ricci curvature bounded from below and positive injectivity radius. Let* $1 \leq q < n$, $1/p = 1/q - 1/n$, *and let* $(f_i)_{i \in I}$ *be functions of class* C^1 *satisfying:*

(i) the f_i's *change sign and there exists* $K > 0$ *such that* $|\nabla f_i| \leq K$ *for any* $i \in I$

(ii) $(|f_i|^q)_{i \in I}$ *is a partition of unity subordinate to a uniformly locally finite covering of M by bounded open sets.*

Then for any $\epsilon > 0$ there exists $B \in \mathbf{R}$ such that

$$\left(\int_M |u|^p dv(g)\right)^{q/p} \leq \left(\frac{K(n,q)^q}{2^{q/n}} + \epsilon\right) \int_M |\nabla u|^q dv(g) + B \int_M |u|^q dv(g)$$

for any $u \in H_1^q(M)$ satisfying: $\forall i \in I$, $\int_M f_i |f_i|^{p-1} |u|^p dv(g) = 0$.

Proof of theorem 4.29: We proceed as in Aubin [Au5]. For $f : M \to \mathbf{R}$ set $f_+ = sup(f, 0)$ and $f_- = sup(-f, 0)$ so that $f = f_+ - f_-$. If $u \in H_1^q(M)$ satisfies the orthogonality conditions of theorem 4.29, then, for any i,

$$\int_M (f_{i+})^p |u|^p dv(g) = \int_M (f_{i-})^p |u|^p dv(g)$$

Independently, $f_{i+}u$ as well as $f_{i-}u$ belong to $H_1^q(M)$. By theorem 4.6 we then get that for any $\epsilon > 0$ there exists $B' \in \mathbf{R}$ such that for any $i \in I$ and any $u \in H_1^q(M)$,

$$\left(\int_M |f_{i\pm}u|^p dv(g)\right)^{q/p} \leq \left(K(n,q)^q + \epsilon\right) \int_M |\nabla(f_{i\pm}u)|^q dv(g) + B' \int_M |f_{i\pm}u|^q dv(g)$$

Suppose now that u satisfies the orthogonality conditions of theorem 4.29 and that

$$\int_M |\nabla(f_{i+}u)|^q dv(g) \geq \int_M |\nabla(f_{i-}u)|^q dv(g)$$

Then,

$$\left(\int_M |f_i u|^p dv(g)\right)^{q/p}$$

$$= 2^{q/p} \left(\int_M |f_{i-}u|^p dv(g)\right)^{q/p}$$

$$\leq 2^{q/p} \left(K(n,q)^q + \epsilon\right) \int_M |\nabla(f_{i-}u)|^q dv(g) + 2^{q/p} B' \int_M |f_{i-}u|^q dv(g)$$

$$\leq 2^{q/p-1} \left(K(n,q)^q + \epsilon\right) \left(\int_M |\nabla(f_{i-}u)|^q dv(g) + \int_M |\nabla(f_{i+}u)|^q dv(g)\right)$$

$$+ 2^{q/p} B' \int_M |f_i u|^q dv(g)$$

$$\leq 2^{-q/n} \left(K(n,q)^q + \epsilon\right) \int_M |\nabla(f_i u)|^q dv(g) + 2^{q/p} B' \int_M |f_i u|^q dv(g)$$

since $1/p = 1/q - 1/n$ and $|\nabla(f_{i-}u)|^q + |\nabla(f_{i+}u)|^q = |\nabla(f_i u)|^q$ a.e. Noting that the result would have been the same under the assumption

$$\int_M |\nabla(f_{i-}u)|^q dv(g) \geq \int_M |\nabla(f_{i+}u)|^q dv(g)$$

we get that for any $\epsilon > 0$ there exists $B'' \in \mathbf{R}$ such that for any $i \in I$ and any $u \in H_1^q(M)$ satisfying the orthogonality conditions of theorem 4.29,

$$\left(\int_M |f_i u|^p dv(g) \right)^{q/p} \leq \left(\frac{K(n,q)^q}{2^{q/n}} + \epsilon \right) \int_M |\nabla(f_i u)|^q dv(g) + B'' \int_M |f_i u|^q dv(g)$$

A similar computation to the one we made in the proof of theorem 4.6 (with $\eta_i = |f_i|^q$) ends the proof of the theorem.

Theorem 4.30: *Let (M,g) be a complete Riemannian n-manifold such that $Rm_{(M,g)}$ and $\nabla Rm_{(M,g)}$ are bounded, and such that $inj_{(M,g)} > 0$. Let $(f_i)_{i \in I}$ be functions of class C^1 satisfying:*

(i) the f_i's change sign and there exists $K > 0$ such that $|\nabla f_i| \leq K$ for any $i \in I$

(ii) $(f_i^2)_{i \in I}$ is a partition of unity subordinate to a uniformly locally finite covering of M by bounded open sets.

Then there exists $B \in \mathbf{R}$ such that

$$\left(\int_M |u|^{2n/(n-2)} dv(g) \right)^{(n-2)/n} \leq \frac{K(n,2)^2}{2^{2/n}} \int_M |\nabla u|^2 dv(g) + B \int_M u^2 dv(g)$$

for any $u \in H_1^2(M)$ satisfying: $\forall i \in I$, $\int_M f_i |f_i|^{(n+2)/(n-2)} |u|^{2n/(n-2)} dv(g) = 0$.

Corollary 4.31: *Let (M,g) be a compact Riemannian n-manifold and let f_i, $i = 1, \ldots, N$, be N functions of class C^1 satisfying $\sum_{i=1}^N f_i^2 = 1$. There exists $B \in \mathbf{R}$ such that*

$$\left(\int_M |u|^{2n/(n-2)} dv(g) \right)^{(n-2)/n} \leq \frac{K(n,2)^2}{2^{2/n}} \int_M |\nabla u|^2 dv(g) + B \int_M u^2 dv(g)$$

for any $u \in H_1^2(M)$ satisfying: $\forall i = 1, \ldots, N$,

$$\int_M f_i |f_i|^{(n+2)/(n-2)} |u|^{2n/(n-2)} dv(g) = 0$$

Proof of theorem 4.30: We first proceed as in the proof of theorem 4.29, using theorem 4.12 instead of theorem 4.6. We then get that there exists $B' \in \mathbf{R}$ such that for any $i \in I$ and any $u \in H_1^2(M)$ satisfying the orthogonality conditions of theorem 4.30,

$$\left(\int_M |f_i u|^{2n/(n-2)} dv(g) \right)^{(n-2)/n}$$

$$\leq \frac{K(n,2)^2}{2^{2/n}} \int_M |\nabla(f_i u)|^2 dv(g) + B' \int_M (f_i u)^2 dv(g)$$

As a consequence,

$$\left(\int_M |u|^{2n/(n-2)} dv(g) \right)^{(n-2)/n}$$

$$\leq \frac{K(n,2)^2}{2^{2/n}} \sum_{i \in I} \int_M |\nabla(f_i u)|^2 dv(g) + B' \sum_{i \in I} \int_M (f_i u)^2 dv(g)$$

But

$$|\nabla(f_i u)|^2 = f_i^2 |\nabla u|^2 + u^2 |\nabla f_i|^2 + u \nabla^\nu u \nabla_\nu f_i^2$$

and since $\sum_{i \in I} f_i^2 = 1$, we get that for any $u \in H_1^2(M)$ satisfying the orthogonality conditions of theorem 4.30,

$$\left(\int_M |u|^{2n/(n-2)} dv(g) \right)^{(n-2)/n}$$

$$\leq \frac{K(n,2)^2}{2^{2/n}} \int_M |\nabla u|^2 dv(g) + (B' + NK^2) \int_M u^2 dv(g)$$

where K is as in theorem 4.30 and where N is such that for any $x \in M$, at most N of the f_i's are nonzero in an open neighborhood of x. This ends the proof of the theorem.

Finally, we mention that important variants of these results exist. This is the case for the following result of Aubin [Au5].

Theorem 4.32: *Let (S^n, h) be the standard unit sphere of $\mathbf{R}^{n+1}$ and let ξ_i, $i = 1, \ldots, n+1$, be the first spherical harmonics obtained by restricting the coordinates x_i of $\mathbf{R}^{n+1}$ to S^n. Let $\epsilon > 0$ and let $1 \leq q < n$, $1/p = 1/q - 1/n$. There exists $B \in \mathbf{R}$ such that*

$$\left(\int_{S^n} |u|^p dv(h) \right)^{q/p} \leq \left(\frac{K(n,q)^q}{2^{q/n}} + \epsilon \right) \int_{S^n} |\nabla u|^q dv(h) + B \int_{S^n} |u|^q dv(h)$$

for any $u \in H_1^q(S^n)$ satisfying: $\forall i = 1, \ldots, n+1$, $\int_{S^n} \xi_i |u|^p dv(h) = 0$.

Chapter 5

Sobolev spaces in the presence of symmetries

The general idea in this chapter is to show that Sobolev embeddings can be improved for functions which possess symmetries. In the first two paragraphs, this includes embeddings in higher L^p spaces and the improvement of the value of the best constant α_q for compact manifolds. The results presented in these two paragraphs are unpublished results of Hebey-Vaugon. In the third paragraph, following Lions [Lio1], we will say some words about the influence of symmetries on the compactness of the Sobolev embeddings.

5.1 EMBEDDINGS IN HIGHER L^p SPACES

First we recall some basic facts concerning the action of Lie groups on manifolds and the space of isometries of a Riemannian manifold. For more details on the following lemmas we refer to Bredon [Br], Dieudonné [Di, chapter 16], and Kobayashi [Ko]. Lemmas 5.2 and 5.3 are respectively due to E.Cartan and Myers-Steenrod [MyS].

Lemma 5.1: *Let G be a compact Lie group acting differentiably on a smooth manifold M. Then for any $x \in M$, $O_G(x) = \{\sigma(x), \sigma \in G\}$ is a smooth compact submanifold of M. Moreover, for any $x \in M$, $S_x = \{\sigma \in G \text{ s.t. } \sigma(x) = x\}$ is a sub-Lie group of G, the quotient manifold G/S_x exists, and the canonical map $\Phi_x : G/S_x \to O_G(x)$ is a diffeomorphism.*

Lemma 5.2: *A closed subgroup of a Lie group is a Lie group.*

Lemma 5.3: *The group $Isom_{(M,g)}$ of isometries of a Riemannian manifold (M,g) is a Lie group with respect to the compact open topology. It acts differentiably on M and if M is compact, $Isom_{(M,g)}$ is also compact.*

From now on, let (M,g) be a compact Riemannian n-manifold, and let G be a subgroup of $Isom_{(M,g)}$. For $x \in M$ we set

$$O_G(x) = \{\sigma(x), \sigma \in G\}$$

and for $p \geq 1$ a real number we set

$$H^p_{1,G}(M) = \{u \in H^p_1(M) \text{ s.t. for any } \sigma \in G, \ u \circ \sigma = u \text{ a.e.}\}$$

$$L_G^p(M) = \{u \in L^p(M) \text{ s.t. for any } \sigma \in G, \ u \circ \sigma = u \text{ a.e.}\}$$

$$C_G^\infty(M) = \{u \in C^\infty(M) \text{ s.t. for any } \sigma \in G, \ u \circ \sigma = u\}$$

One then has the following.

Lemma 5.4: *Let (M, g) be a compact Riemannian manifold and let G be a subgroup of $Isom_{(M,g)}$. For any real number $q \geq 1$, $C_G^\infty(M)$ is dense in $H_{1,G}^q(M)$.*

Proof of lemma 5.4: Let $u \in H_{1,G}^q(M)$ and let (u_m) be a sequence in $C^\infty(M)$ such that $\lim\limits_{m \to +\infty} u_m = u$ in $H_1^q(M)$. If $\overline{G}$ denotes the closure of G in $Isom_{(M,g)}$, then, clearly, $u \circ \sigma = u$ a.e for any $\sigma \in \overline{G}$. Let $d\sigma$ be the Haar measure on $\overline{G}$ and set

$$\tilde{u}_m(x) = \frac{1}{\int_{\overline{G}} d\sigma} \int_{\overline{G}} u_m(\sigma(x)) d\sigma \ ,$$

$x \in M$. One easily checks that for any m, $\tilde{u}_m \in C_G^\infty(M)$, and that $\lim\limits_{m \to +\infty} \tilde{u}_m = u$ in $H_1^q(M)$. This ends the proof of the lemma.

Let G and (M, g) be as above, and let $\overline{G}$ be the closure of G in $Isom_{(M,g)}$. One easily sees that

(i) if $u : M \to \mathbf{R}$ is G-invariant, then u is also $\overline{G}$-invariant

(ii) $\overline{G}$ has a finite number of connected components, and if G_1 and G_2 are two such components there exists $\sigma \in \overline{G}$ such that $\sigma G_1 = G_2$

(iii) if G_0 denotes the connected component of the identity in $\overline{G}$, G_0 is a compact Lie group.

In particular, if for some $x \in M$, $CardO_G(x) = +\infty$, then by (ii) one also has that $CardO_{G_0}(x) = +\infty$ (where $Card$ stands for the cardinality). Let us now prove the following. (Note that by lemma 5.1, $O_{G_0}(x)$ is a smooth connected submanifold of M).

Lemma 5.5: *Let (M, g) be a compact Riemannian n-manifold, let G be a subgroup of $Isom_{(M,g)}$, and let G_0 be the connected component of the identity in $\overline{G}$ (the closure of G in $Isom_{(M,g)}$). Let $x \in M$ and set $k = dimO_{G_0}(x)$. Assume $k \geq 1$. There exists a coordinate chart (Ω, ϕ) of M at x such that:*

(1) $\phi(\Omega) = U \times V$, where U is some open subset of $\mathbf{R}^k$ and V is some open subset of $\mathbf{R}^{n-k}$

(2) $\forall y \in \Omega, \ U \times \Pi_2(\phi(y)) \subset \phi(O_{G_0}(y) \cap \Omega)$ where $\Pi_2 : \mathbf{R}^k \times \mathbf{R}^{n-k} \to \mathbf{R}^{n-k}$ is the second projection.

Proof of lemma 5.5: Let $S_x = \{\sigma \in G_0 \text{ s.t. } \sigma(x) = x\}$ and let $\Phi : G_0 \to M$ be defined by $\Phi(\sigma) = \sigma(x)$. A classical result states that Φ has constant rank (see for instance [Di, chapter 16]). Since $S_x = \Phi^{-1}(x)$, we get that

$$dimS_x = dimG_0 - Rank\Phi$$

On the other hand, by lemma 5.1,

$$dim(G_0/S_x) = dimO_{G_0}(x) = dimG_0 - dimS_x$$

Hence, $Rank\Phi = k$. As a consequence, there exists a k-dimensional submanifold H of G_0 such that $Id \in H$ and $\Phi_{|H}$ is an embedding. Let N be a $(n-k)$-dimensional submanifold of M such that

$$T_x\Phi(H) \oplus T_xN = T_xM$$

and let $\Psi : H \times N \to M$ be defined by $\Psi(\sigma, y) = \sigma(y)$. Clearly, Ψ is smooth and $D\Psi_{(Id,x)}$ is an isomorphism. Let (U', ϕ_1) be a chart of H at Id and (V', ϕ_2) be a chart of N at x, U' and V' being such that $\Psi_{|U'\times V'}$ is a diffeomorphism. To get the lemma one just has to set $\Omega = \Psi(U' \times V')$ and $\phi = (\phi_1 \circ \Psi_1^{-1}, \phi_2 \circ \Psi_2^{-1})$, where $\Psi^{-1} = (\Psi_1^{-1}, \Psi_2^{-1})$.

With such a lemma we are now in position to prove our first result. Roughly speaking, it shows that for functions which possess enough symmetries, the Sobolev embeddings are valid in higher L^p spaces. Similar results have been obtained in specific contexts by Cotsiolis-Iliopoulos [CoI], Lions [Lio2], and Ding [Din].

Theorem 5.6: *Let (M, g) be a compact Riemannian n-manifold, G a subgroup of $Isom_{(M,g)}$, and $q \geq 1$ a real number. Assume that for any $x \in M$, $CardO_G(x) = +\infty$. Let $k = min_{x\in M} dimO_{G_0}(x)$ where G_0 is the connected component of the identity in $\overline{G}$ (the closure of G in $Isom_{(M,g)}$). Then:*

(1) if $n - k \leq q$, for any real number $p \geq 1$, $H^q_{1,G}(M) \subset L^p(M)$ and these embeddings are compact

(2) if $n - k > q$, for any real number $1 \leq p \leq \frac{(n-k)q}{(n-k-q)}$, $H^q_{1,G}(M) \subset L^p(M)$ and these embeddings are compact provided that $p < \frac{(n-k)q}{(n-k-q)}$.

Remark: Recall that the assumption that for any $x \in M$, $CardO_G(x) = +\infty$, implies that for any $x \in M$, $CardO_{G_0}(x) = +\infty$. Independently, note that by lemma 5.1, for any $x \in M$, $O_{G_0}(x)$ is a connected smooth submanifold of M. Hence, $k \geq 1$. One then easily checks that for $q < n - k$, $\frac{(n-k)q}{(n-k-q)} > \frac{nq}{(n-q)}$.

Corollary 5.7: *Let (M,g) be a compact Riemannian n-manifold, and let G be a subgroup of $Isom_{(M,g)}$. Assume that for any $x \in M$, $CardO_G(x) = +\infty$. Then for any $1 \leq q < n$ there exists $p_0 > nq/(n-q)$ such that for any $1 \leq p \leq p_0$ the embedding of $H^q_{1,G}(M)$ in $L^p(M)$ is valid and compact.*

Proof of theorem 5.6: By lemma 5.5 we can assume that M is covered by a finite number of charts $(\Omega_m, \phi_m)_{m=1,...,N}$ satisfying for any m:

(i) $\phi_m(\Omega_m) = U_m \times V_m$, where U_m is some open subset of $\mathbf{R}^{k_m}$, V_m is some open subset of $\mathbf{R}^{n-k_m}$, and $k_m \in \mathbf{N}$ satsifies $k_m \geq k$

(ii) U_m, V_m are bounded, and V_m has smooth boundary

(iii) $\forall y \in \Omega_m$, $U_m \times \Pi_2(\phi_m(y)) \subset \phi_m(O_{G_0}(y) \cap \Omega_m)$

(iv) $\exists \alpha_m > 0$ with $\alpha_m^{-1}\delta_{ij} \leq g^m_{ij} \leq \alpha_m \delta_{ij}$ as bilinear forms

where in (iii), $\Pi_2 : \mathbf{R}^{k_m} \times \mathbf{R}^{n-k_m} \to \mathbf{R}^{n-k_m}$ denotes the second projection, and in (iv), the g^m_{ij}'s denote the components of g in (Ω_m, ϕ_m). Let $u \in C^\infty_G(M)$. Since $u \circ \sigma = u$ for any $\sigma \in G_0$ (as one easily checks), we get that for any m, any $x, x' \in U_m$, and any $y \in V_m$, $u \circ \phi_m^{-1}(x,y) = u \circ \phi_m^{-1}(x',y)$. As a consequence, for any m there exists $\tilde{u}_m \in C^\infty(\mathbf{R}^{n-k_m}, \mathbf{R})$ such that for any $x \in U_m$ and any $y \in V_m$, one has that $u \circ \phi_m^{-1}(x,y) = \tilde{u}_m(y)$. (Without loss of generality, one can assume that ϕ_m is actually defined on some open set $\tilde{\Omega}_m$ containing $\overline{\Omega}_m$, and such that $\phi_m(\tilde{\Omega}_m) = \tilde{U}_m \times \tilde{V}_m$ with $\overline{V}_m \subset \tilde{V}_m$). We then get that for any m and any real number $p \geq 1$,

$$\int_{\Omega_m} |u|^p dv(g) = \int_{U_m \times V_m} \left(|u|^p \sqrt{\det g^m_{ij}} \right) \circ \phi_m^{-1}(x,y) dx dy$$

$$\leq A_m \int_{U_m \times V_m} |u \circ \phi_m^{-1}(x,y)|^p dx dy$$

$$= \tilde{A}_m \int_{V_m} |\tilde{u}_m(y)|^p dy$$

where A_m and $\tilde{A}_m$ are positive constants which do not depend on u. Similarly, one has that for any m and any $p \geq 1$,

$$\int_{\Omega_m} |u|^p dv(g) \geq B_m \int_{V_m} |\tilde{u}_m(y)|^p dy$$

and

$$\int_{\Omega_m} |\nabla u|^p dv(g) \geq \tilde{B}_m \int_{V_m} |\nabla \tilde{u}_m(y)|^p dy$$

where $B_m > 0$ and $\tilde{B}_m > 0$ do not depend on u. Combining these inequalities and the Sobolev embedding theorem for bounded domains of euclidean spaces (see for instance [Ad, theorem 5.4]), we get that for any m and any real number $q \geq 1$,

(v) if $n - k_m \leq q$, then for any real number $p \geq 1$ there exists $C_m > 0$ such that for any $u \in C_G^\infty(M)$,

$$\left(\int_{\Omega_m} |u|^p \, dv(g)\right)^{1/p} \leq C_m \left(\left(\int_{\Omega_m} |\nabla u|^q \, dv(g)\right)^{1/q} + \left(\int_{\Omega_m} |u|^q \, dv(g)\right)^{1/q}\right)$$

(vi) if $n - k_m > q$, then for any real number $1 \leq p \leq (n - k_m)q/(n - k_m - q)$ there exists $C_m > 0$ such that for any $u \in C_G^\infty(M)$,

$$\left(\int_{\Omega_m} |u|^p \, dv(g)\right)^{1/p} \leq C_m \left(\left(\int_{\Omega_m} |\nabla u|^q \, dv(g)\right)^{1/q} + \left(\int_{\Omega_m} |u|^q \, dv(g)\right)^{1/q}\right)$$

But:

(vii) $n - k_m \leq n - k$ so that for $q < n - k_m$, $\frac{(n-k_m)q}{(n-k_m-q)} \geq \frac{(n-k)q}{(n-k-q)}$

(viii) $\left(\int_M |u|^p \, dv(g)\right)^{1/p} \leq \sum_{m=1}^N \left(\int_{\Omega_m} |u|^p \, dv(g)\right)^{1/p}$

(ix) $\sum_{m=1}^N \left(\left(\int_{\Omega_m} |\nabla u|^q \, dv(g)\right)^{1/q} + \left(\int_{\Omega_m} |u|^q \, dv(g)\right)^{1/q}\right)$

$$\leq N \left(\left(\int_M |\nabla u|^q \, dv(g)\right)^{1/q} + \left(\int_M |u|^q \, dv(g)\right)^{1/q}\right)$$

As a consequence, for any real number $q \geq 1$:

(x) if $n - k \leq q$, then for any real number $p \geq 1$, $H_{1,G}^q(M) \subset L^p(M)$

(xi) if $n - k > q$, then for any real number $1 \leq p \leq (n - k)q/(n - k - q)$, $H_{1,G}^q(M) \subset L^p(M)$

Independently, by standard arguments and [Ad, theorem 6.2], one easily gets that these embeddings are compact for any $p \geq 1$ in case (x), and any $p < \frac{(n-k)q}{(n-k-q)}$ in case (xi). This ends the proof of the theorem.

Remarks: 1) We decided in this section to work with G_0 instead of $\overline{G}$. There is no particular reason for that, apart from the fact that $O_{G_0}(x)$ for $x \in M$ is a "nice" connected submanifold of M. On the other hand, $O_{\overline{G}}(x)$ is also a submanifold of M, and since $\overline{G}$ has a finite number of connected components, clearly $dimO_{\overline{G}}(x) = dimO_{G_0}(x)$ for any $x \in M$. As a consequence, one can set $k = min_{x \in M} dimO_{\overline{G}}(x)$ in theorem 5.6.

2) Embeddings in higher L^p spaces for non compact manifolds have recently been obtained by Hebey and Vaugon [HV5]. While one just has to consider the minimum orbit dimension of G for compact manifolds, it turns out that when dealing with non compact manifolds one has also to consider the "geometry" of the action of G at infinity. As an example, and when the action of $G \subset Isom_{(M,g)}$ is of codimension 1, G compact, one gets continuous embeddings in the spirit of theorem 5.6 if there exists a compact subset K of M and a positive constant C such that for any $x \in M \backslash K$, $v(O_G(x)) \geq C$, where $v(O_G(x))$ denotes the volume of $O_G(x)$ for the metric induced by g. In the same order of ideas, one gets compact embeddings in the spirit of theorem 5.6 if for any $\epsilon > 0$ there exists a compact subset K_ϵ of M such that for any $x \in M \backslash K_\epsilon$, $v(O_G(x)) \geq 1/\epsilon$. We refer the reader to [HV5] for more details on the subject.

5.2 IMPROVEMENT OF THE BEST CONSTANTS

When G has finite orbits, results such as corollary 5.7 are obviously false. In this case the interesting question is to know if one can lowered the value of the best constant α_q. The answer to this question is given by theorems 5.8 and 5.10 below.

Theorem 5.8: *Let (M,g) be a compact Riemannian n-manifold and let G be a subgroup of $Isom_{(M,g)}$. Let $k = inf_{x \in M} CardO_G(x)$. For any $1 \leq q < n$ and any $\epsilon > 0$ there exists $B \in \mathbf{R}$ such that for any $u \in H_{1,G}^q(M)$,*

$$\left(\int_M |u|^p dv(g) \right)^{q/p} \leq \left(\frac{K(n,q)^q}{k^{q/n}} + \epsilon \right) \int_M |\nabla u|^q dv(g) + B \int_M |u|^q dv(g)$$

where $1/p = 1/q - 1/n$, $K(n,q)$ is as in theorem 4.4, and $\frac{K(n,q)^q}{k^{q/n}} = 0$ if $k = +\infty$.

In order to prove theorem 5.8, we need the following lemma.

Lemma 5.9: *Let (M,g) be a compact Riemannian n-manifold, G a subgroup of $Isom_{(M,g)}$, G_0 the connected component of the identity in $\overline{G}$ (the closure of G in $Isom_{(M,g)}$), and p, q two real numbers such that $1 \leq q < n$ and $p = nq/(n-q)$. Let O be a compact subset of M such that O is stable under the action of G_0 (i.e $\sigma O = O$, for any $\sigma \in G_0$), and such that for any $x \in O$, $CardO_{G_0}(x) = +\infty$. Then there exists $\delta > 0$ such that for any $\epsilon > 0$ there exists $B \in \mathbf{R}$ with the following property: for any $u \in C_G^\infty(M)$ satisfying*

$$Suppu \subset O_\delta = \{y \in M \text{ s.t. } d_g(y,O) \leq \delta\},$$

one has that

$$\left(\int_M |u|^p \, dv(g)\right)^{q/p} \leq \epsilon \int_M |\nabla u|^q \, dv(g) + B \int_M |u|^q \, dv(g)$$

Proof of lemma 5.9: Since O is compact, O is covered by a finite number of charts (Ω_m, ϕ_m) satisfying the assumptions (i) to (iv) of the proof of theorem 5.6 (with $k \geq 1$ given by $k = min_{x \in O} dim O_{G_0}(x)$). We choose $\delta > 0$ such that $O_\delta \subset \cup \Omega_m$. Let $1 \leq q < n$ and $p = nq/(n-q)$. Set

$$H_q(M) = \left\{u \in H_{1,G}^q(M) \text{ s.t. } Supp u \subset O_\delta\right\}$$

Similar arguments to those developed in the proof of theorem 5.6 will prove that the embedding of $H_q(M)$ in $L^p(M)$ is compact. Independently, by a classical result of Lions [Lio], if $\mathcal{B}_1$, $\mathcal{B}_2$, $\mathcal{B}_3$ are three Banach spaces, $u : \mathcal{B}_1 \to \mathcal{B}_2$ is a compact linear operator, and $v : \mathcal{B}_2 \to \mathcal{B}_3$ is a continuous one to one linear operator, then, for any $\epsilon > 0$ there exists $B > 0$ such that for any $x \in \mathcal{B}_1$,

$$\|u(x)\|_{\mathcal{B}_2} \leq \epsilon \|x\|_{\mathcal{B}_1} + B \|v \circ u(x)\|_{\mathcal{B}_3}$$

Applying this result with $\mathcal{B}_1 = H_q(M)$, $\mathcal{B}_2 = L^p(M)$, and $\mathcal{B}_3 = L^q(M)$, one gets the lemma.

Let us now prove theorem 5.8.

Proof of theorem 5.8: If $k = +\infty$, theorem 5.8 is an easy consequence of corollary 5.7 and the result of Lions mentioned above. One can then suppose that $k < +\infty$. Let $1 \leq q < n$ be given, $\overline{G}$ be the closure of G in $Isom_{(M,g)}$, and G_0 be the connected component of the identity in $\overline{G}$. Clearly, for any $x \in M$, $Card O_{\overline{G}}(x) = Card O_G(x)$. Let $x \in M$.

If $Card O_{\overline{G}}(x) < +\infty$, let $O_{\overline{G}}(x) = \{x_1, \ldots, x_m\}$ with the convention that $x_1 = x$. We then choose $\delta = \delta(x) \in (0, inj_{(M,g)})$ small enough such that for any $i \neq j$, $B_{x_i}(\delta) \cap B_{x_j}(\delta) = \emptyset$, and we note $U_x = \cup_{j=1}^m B_{x_j}(\delta)$.

Suppose now that $Card O_{\overline{G}}(x) = +\infty$. By lemma 5.1, and since $\overline{G}$ has a finite number of connected components, $O_{\overline{G}}(x)$ is a smooth compact submanifold of M of dimension greater than or equal to 1. Let $O_{\overline{G}}(x) = O_1 \cup \ldots \cup O_m$, with the convention that $x \in O_1$, the O_i's being the connected components of $O_{\overline{G}}(x)$. The O_i's are compact since $O_{\overline{G}}(x)$ is compact. Furthermore, O_1 is clearly stable under the action of G_0 and for any $y \in O_1$, one has that $Card O_{G_0}(y) = +\infty$. We now choose $\delta = \delta(x)$ small enough such that

(i) δ is less than the one given by lemma 5.9 (with $O = O_1$)

(ii) for any $i \neq j$, $B_i^\delta \cap B_j^\delta = \emptyset$ where $B_i^\delta = \{y \in M \text{ s.t. } d_g(y, O_i) \leq \delta\}$

(iii) for any i, $\phi_i : \overset{\circ}{B_i^{2\delta}} \to \mathbf{R}$ defined by $\phi_i(y) = d_g(y, O_i)^2$ is smooth.

Here again, we note $U_x = \cup_{j=1}^m B_j^\delta$. One then has that for any $x \in M$, U_x is stable under the action of $\overline{G}$. Now, since M is compact, let $x_1, \ldots, x_N \in M$ be such that $M = \cup_{i=1}^N \overset{\circ}{U}_{x_i}$. For any $\epsilon > 0$, let $f_\epsilon \in C^\infty(\mathbf{R}, \mathbf{R})$ be such that $f_\epsilon(t) > 0$ if $t < \epsilon$ and $f_\epsilon(t) = 0$ if $t \geq 0$. For any $i = 1, \ldots, N$ we set

$$\alpha_{ij}(x) = f_{\delta_i}\left(d_g(x, x_{ij})^2\right) \quad \text{if} \quad U_{x_i} = \cup_{j=1}^{m_i} B_{x_{ij}}(\delta_i), \ \delta_i = \delta(x_i)$$

$$\alpha_{ij}(x) = f_{\delta_i}\left(d_g(x, O_{ij})^2\right) \quad \text{if} \quad U_{x_i} = \cup_{j=1}^{m_i} B_j^{\delta_i}, \ \delta_i = \delta(x_i)$$

The α_{ij}'s, $i = 1, \ldots, N$, $j = 1, \ldots, m_i$, are smooth functions and

$$\eta_{ij} = \frac{\alpha_{ij}^{[q]+1}}{\sum_{\mu,\nu} \alpha_{\mu\nu}^{[q]+1}}$$

$i = 1, \ldots, N$, $j = 1, \ldots, m_i$, is a smooth partition of unity of M which satisfies:

(iv) for any i, j, $\eta_{ij}^{1/q} \in C^1(M)$

(v) there exists $H \in \mathbf{R}$ such that for any i, j, $|\nabla(\eta_{ij}^{1/q})| \leq H$

(vi) for any $i = 1, \ldots, N$ and any $j \neq j' = 1, \ldots, m_i$, $\eta_{ij}\eta_{ij'} = 0$

Furthermore, one easily sees that for any $i = 1, \ldots, N$ and any $j, j' = 1, \ldots, m_i$, there exists $\sigma \in \overline{G}$ such that $\eta_{ij'} = \eta_{ij} \circ \sigma$. According to what we have just said one then has that for any $u \in C_G^\infty(M)$,

$$\left(\int_M |u|^p dv(g)\right)^{q/p} = \left(\int_M |\sum_{i,j} \eta_{ij}|u|^q|^{p/q} dv(g)\right)^{q/p}$$

$$\leq \sum_{i=1}^N \left(\int_M |\sum_{j=1}^{m_i} \eta_{ij}|u|^q|^{p/q} dv(g)\right)^{q/p}$$

$$= \sum_{i=1}^N m_i^{q/p} \left(\int_M |\eta_{i1}^{1/q} u|^p dv(g)\right)^{q/p}$$

Let $i \in \{1, \ldots, N\}$ be given and suppose that $CardO_G(x_i) < +\infty$. By theorem 4.6, for any $\epsilon_i > 0$ there exists $B_i \in \mathbf{R}$ such that for any $u \in C_G^\infty(M)$,

$$\left(\int_M |\eta_{i1}^{1/q} u|^p dv(g)\right)^{q/p}$$

$$\leq \left(K(n,q)^q + \epsilon_i\right) \int_M |\nabla(\eta_{i1}^{1/q} u)|^q dv(g) + B_i \int_M \eta_{i1}|u|^q dv(g)$$

Independently, suppose that $CardO_G(x_i) = +\infty$. Since η_{i1} is G_0-invariant (as one easily checks), we get by lemma 5.9 that for any $\epsilon_i > 0$ there exists $B_i \in \mathbf{R}$ such that for any $u \in C_G^\infty(M)$,

$$\left(\int_M |\eta_{i1}^{1/q} u|^p dv(g) \right)^{q/p} \leq \epsilon_i \int_M |\nabla(\eta_{i1}^{1/q} u)|^q dv(g) + B_i \int_M \eta_{i1} |u|^q dv(g)$$

Let $I_1 = \{i \text{ s.t. } CardO_G(x_i) < +\infty\}$ and $I_2 = \{i \text{ s.t. } CardO_G(x_i) = +\infty\}$. With a similar computation to the one we made in the proof of theorem 4.6 we then get that for any $u \in C_G^\infty(M)$,

$$\|u\|_p^q \leq \sum_{i \in I_1} m_i^{q/p} \left(K(n,q)^q + \epsilon_i \right) \left(\int_M |\nabla u|^q \eta_{i1} dv(g) \right.$$

$$+ \mu H \|\nabla u\|_q^{q-1} \|u\|_q + \nu H^q \|u\|_q^q \bigg)$$

$$+ \sum_{i \in I_2} m_i^{q/p} \epsilon_i \left(\int_M |\nabla u|^q \eta_{i1} dv(g) + \mu H \|\nabla u\|_q^{q-1} \|u\|_q \right.$$

$$+ \nu H^q \|u\|_q^q \bigg) + N \left(max_{i=1,\ldots,N} B_i m_i^{q/p} \right) \|u\|_q^q$$

where $\mu > 0$ and $\nu > 0$ depend only on q, and where $\|f\|_p = \left(\int_M |f|^p dv(g) \right)^{1/p}$. Independently, for any $i = 1, \ldots, N$,

$$\int_M |\nabla u|^q \eta_{i1} dv(g) = \frac{1}{m_i} \sum_{j=1}^{m_i} \int_M |\nabla u|^q \eta_{ij} dv(g)$$

while for any $i \in I_1$, $m_i \geq k = min_{x \in M} CardO_G(x)$. Since $1 - q/p = q/n$, if we choose $\epsilon_i = \epsilon$ when $i \in I_1$ and $\epsilon_i \leq K(n,q)^q (m_i/k)^{q/n}$ when $i \in I_2$, we get that for any $\epsilon > 0$ there exists $B \in \mathbf{R}$ such that for any $u \in C_G^\infty(M)$,

$$\|u\|_p^q \leq \frac{K(n,q)^q + \epsilon}{k^{q/n}} \left(\|\nabla u\|_q^q + \mu H \left(\sum_{i=1}^N m_i \right) \|\nabla u\|_q^{q-1} \|u\|_q \right.$$

$$+ \nu H^q \left(\sum_{i=1}^N m_i \right) \|u\|_q^q \bigg) + B \|u\|_q^q$$

Noting that for any $\epsilon > 0$ there exists $C_\epsilon > 0$ such that for any positive real numbers x and y, $x^{q-1} y \leq \epsilon x^q + C_\epsilon y^q$, one easily obtains the inequality of theorem 5.8 from this last inequality. This ends the proof of the theorem.

Theorem 5.10: *Let (M,g) be a compact Riemannian n-manifold and let G be a subgroup of $Isom_{(M,g)}$ which possesses at least one finite orbit. Let $k = min_{x \in M} Card O_G(x)$. There exists $B \in \mathbf{R}$ such that for any $u \in H^2_{1,G}(M)$,*

$$\left(\int_M |u|^{2n/(n-2)} dv(g) \right)^{(n-2)/n} \leq \frac{K(n,2)^2}{k^{2/n}} \int_M |\nabla u|^2 dv(g) + B \int_M u^2 dv(g)$$

where $K(n,2)$ is as in theorem 4.4.

Proof of theorem 5.10: We first proceed as in the proof of theorem 5.8, using theorem 4.12 instead of theorem 4.6. We then have that for any $i \in I_1$ there exists $B_i \in \mathbf{R}$ such that for any $u \in C^\infty_G(M)$,

$$\left(\int_M |\eta_{i1}^{1/2} u|^{2n/(n-2)} dv(g) \right)^{(n-2)/n}$$
$$\leq K(n,2)^2 \int_M |\nabla(\eta_{i1}^{1/2} u)|^2 dv(g) + B_i \int_M \eta_{i1} u^2 dv(g)$$

As a consequence, for any $u \in C^\infty_G(M)$,

$$\left(\int_M |u|^{2n/(n-2)} dv(g) \right)^{(n-2)/n}$$
$$\leq \sum_{i \in I_1} m_i^{(n-2)/n} K(n,2)^2 \left(\int_M |\nabla u|^2 \eta_{i1} dv(g) \right.$$
$$\left. + \int_M u^2 |\nabla(\eta_{i1}^{1/2} u)|^2 dv(g) + \int_M u \nabla^\nu u \nabla_\nu \eta_{i1} dv(g) \right)$$
$$+ \sum_{i \in I_2} m_i^{(n-2)/n} \epsilon_i \left(\int_M |\nabla u|^2 \eta_{i1} dv(g) \right.$$
$$\left. + \int_M u^2 |\nabla(\eta_{i1}^{1/2} u)|^2 dv(g) + \int_M u \nabla^\nu u \nabla_\nu \eta_{i1} dv(g) \right)$$
$$+ N \left(max_{i=1,\dots,N} B_i m_i^{(n-2)/n} \right) \int_M u^2 dv(g)$$

Independently, for any $i = 1, \dots, N$,

$$\int_M |\nabla u|^2 \eta_{i1} dv(g) = \frac{1}{m_i} \sum_{j=1}^{m_i} \int_M |\nabla u|^2 \eta_{ij} dv(g)$$

and

$$\int_M u \nabla^\nu u \nabla_\nu \eta_{i1} dv(g) = \frac{1}{m_i} \sum_{j=1}^{m_i} \int_M u \nabla^\nu u \nabla_\nu \eta_{ij} dv(g)$$

while for any $i \in I_1$, $m_i \geq k = min_{x \in M} Card O_G(x)$. Choosing

$$\epsilon_i \leq K(n,2)^2 (m_i/k)^{2/n}$$

we then get that for any $u \in C_G^\infty(M)$,

$$\left(\int_M |u|^{2n/(n-2)} dv(g) \right)^{(n-2)/n}$$

$$\leq \frac{K(n,2)^2}{k^{2/n}} \sum_{i,j} \int_M |\nabla u|^2 \eta_{ij} dv(g)$$

$$+ \frac{K(n,2)^2}{k^{2/n}} \sum_{i,j} \int_M u \nabla^\nu u \nabla_\nu \eta_{ij} dv(g)$$

$$+ \frac{K(n,2)^2}{k^{2/n}} \left(\sum_{i=1}^N m_i \right) H^2 \int_M u^2 dv(g)$$

$$+ N \left(max_{i=1,...,N} B_i m_i^{(n-2)/n} \right) \int_M u^2 dv(g)$$

$$= \frac{K(n,2)^2}{k^{2/n}} \sum_{i,j} \int_M |\nabla u|^2 \eta_{ij} dv(g)$$

$$+ \frac{K(n,2)^2}{k^{2/n}} \left(\sum_{i=1}^N m_i \right) H^2 \int_M u^2 dv(g)$$

$$+ N \left(max_{i=1,...,N} B_i m_i^{(n-2)/n} \right) \int_M u^2 dv(g)$$

since $\sum_{i,j} \eta_{ij} = 1$. Letting

$$B = \frac{K(n,2)^2}{k^{2/n}} \left(\sum_{i=1}^N m_i \right) H^2 + N \left(max_{i=1,...,N} B_i m_i^{(n-2)/n} \right)$$

this ends the proof of the theorem.

5.3 COMPACTNESS AND SYMMETRIES

We present in this paragraph two results obtained by Lions [Lio1]. In particular, we restrict ourselves to the Euclidean context. The idea is to show that Sobolev embeddings become compact in the presence of symmetries. This phenomenon was first observed by several authors. We refer for instance to Berestycki-Lions [BeL], Coleman-Glazer-Martin [CoGM], and Strauss [St]. The proof of Lions results we present here, due to Vaugon (oral communication), is slightly simpler than the one of [Lio1]. For sake of completeness, we mention that similar results in the Riemannian context have recently been obtained by Hebey-Vaugon [HV5].

For $q \geq 1$ real, set

$$H_{1,r}^q(\mathbf{R}^n) = \{u \in H_1^q(\mathbf{R}^n) \text{ s.t. } u \text{ is radially symmetric}\}$$

By radially symmetric we mean here that the function is invariant under the action of $O(n)$. With similar arguments to those used in the proof of lemma 5.4 one easily gets that

$$\mathcal{D}_r(\mathbf{R}^n) = \{u \in \mathcal{D}(\mathbf{R}^n) \text{ s.t. } u \text{ is radially symmetric}\}$$

is dense in $H_{1,r}^q(\mathbf{R}^n)$. We then have the following.

Theorem 5.11: *For any $1 \leq q < n$ and any $q < p < nq/(n-q)$, the embedding of $H_{1,r}^q(\mathbf{R}^n)$ in $L^p(\mathbf{R}^n)$ is compact.*

Proof of theorem 5.11: Let $1 \leq q < n$ and $q < p < nq/(n-q)$ be given. By the mean value theorem for integrals one easily gets that there exists a positive constant C such that for any $f \in C^\infty([0,1])$,

$$\int_0^1 |f(t)|^p \, dt \leq C \Big(\int_0^1 (|f'(t)|^q + |f(t)|^q) \, dt \Big)^{p/q}$$

It is then easy to see that for any integer k and any $f \in C^\infty([k, k+1])$,

$$\int_k^{k+1} |f(t)|^p \, dt \leq C \Big(\int_k^{k+1} (|f'(t)|^q + |f(t)|^q) \, dt \Big)^{p/q}$$

Let k be an integer and set

$$C_k = \{x \in \mathbf{R}^n \text{ s.t. } k \leq |x| \leq k+1\}$$

Noting that $p/q > 1$ one then has that for any $u \in \mathcal{D}_r(\mathbf{R}^n)$ and any real number $R \geq 1$,

$$\int_{R^n \backslash B_0^c(R)} |u(x)|^p \, dx$$

$$\leq \sum_{k \geq [R]} \int_{C_k} |u(x)|^p \, dx$$

$$\leq \omega_{n-1} \sum_{k \geq [R]} (k+1)^{n-1} \int_k^{k+1} |u(t)|^p \, dt$$

$$\le C\omega_{n-1} \sum_{k\ge[R]} (k+1)^{n-1} \left(\int_k^{k+1} (|u'(t)|^q + |u(t)|^q)\, dt \right)^{p/q}$$

$$= C\omega_{n-1} \sum_{k\ge[R]} (k+1)^{(n-1)(1-p/q)}$$

$$\times \left(\left(\frac{k+1}{k}\right)^{n-1} k^{n-1} \int_k^{k+1} (|u'(t)|^q + |u(t)|^q)\, dt \right)^{p/q}$$

$$\le 2^{(n-1)p/q} C\omega_{n-1}^{1-p/q}$$

$$\times \sum_{k\ge[R]} (k+1)^{(n-1)(1-p/q)} \left(\int_{C_k} (|\nabla u(x)|^q + |u(x)|^q)\, dx \right)^{p/q}$$

$$\le \frac{2^{(n-1)p/q} C\omega_{n-1}^{1-p/q}}{([R]+1)^{(n-1)(p/q-1)}} \left(\int_{R^n\setminus B_0^c(R)} (|\nabla u(x)|^q + |u(x)|^q)\, dx \right)^{p/q}$$

As a consequence, we get that there exists a positive constant A such that for any $R \ge 1$ and any $u \in \mathcal{D}_r(\mathbf{R}^n)$,

$$\left(\int_{R^n\setminus B_0^c(R)} |u(x)|^p\, dx \right)^{1/p} \le A([R]+1)^{-(n-1)(1/q-1/p)} \|u\|_{H_1^q}$$

By density such an inequality is then valid for any $u \in H_{1,r}^q(\mathbf{R}^n)$. Independently, since $1/q - 1/p > 0$, one has that

$$\lim_{R\to+\infty} ([R]+1)^{-(n-1)(1/q-1/p)} = 0$$

By [Ad, theorem 2.22] this ends the proof of the theorem.

From now on, let $m \ge 2$ and let $n = \sum_{i=1}^m n_i$ with $n_i \ge 2$. (m, n, and n_i being integers). For $q \ge 1$ real set

$$H_{1,s}^q(\mathbf{R}^n) = \left\{ u \in H_1^q(\mathbf{R}^n) \text{ s.t. for any } i = 1,\ldots,m,\, u \text{ is radially} \right.$$
$$\left. \text{symmetric with respect to } x_i \in \mathbf{R}^{n_i} \right\}$$

We mean here that for any $(x_1^0,\ldots,x_m^0) \in \mathbf{R}^{n_1} \times \ldots \times \mathbf{R}^{n_m}$, and any integer $i \in \{1,\ldots,m\}$, the function

$$x_i \to u(x_1^0,\ldots,x_{i-1}^0, x_i, x_{i+1}^0,\ldots,x_m^0)$$

is radially symmetric (i.e invariant under the action of $O(n_i)$). Functions having such symmetries are said to be cylindrically symmetric. Here again, similar arguments to those used in the proof of lemma 5.4 show that

$$\mathcal{D}_s(\mathbf{R}^n) = \left\{ u \in \mathcal{D}(\mathbf{R}^n) \text{ s.t. } u \text{ is cylindrically symmetric} \right\}$$

is dense in $H_{1,s}^q(\mathbf{R}^n)$. One then has the following.

Theorem 5.12: *For any $1 \leq q < n$ and any $q < p < nq/(n-q)$, the embedding of $H_{1,s}^q(\mathbf{R}^n)$ in $L^p(\mathbf{R}^n)$ is compact.*

Proof of theorem 5.12: For sake of simplicity we suppose that $m = 2$. Let (a_k) be a sequence of points in $\mathbf{R}^{n_1}$ such that $(B_{a_k}^1(1))$ is a uniformly locally finite covering of $\mathbf{R}^{n_1}$, where $B_{a_k}^1(1)$ is the unit ball of center a_k in $\mathbf{R}^{n_1}$, and let (b_k) be a sequence of points of $\mathbf{R}^{n_2}$ such that $(B_{b_k}^2(1))$ is a uniformly locally finite covering of $\mathbf{R}^{n_2}$, where $B_{b_k}^2(1)$ is the unit ball of center b_k in $\mathbf{R}^{n_2}$. If $R \geq 1$ is given, we set

$$V_\alpha = B_{a_i}^1(1) \times C_j^2, \quad i \in \mathbf{N}, \; j \geq [R]$$
$$W_\beta = C_i^1 \times B_{b_j}^2(1), \quad i \geq [R], \; j \in \mathbf{N}$$

where

$$C_i^1 = \{x \in \mathbf{R}^{n_1} \text{ s.t. } i \leq |x| \leq i+1\}$$
$$C_j^2 = \{y \in \mathbf{R}^{n_2} \text{ s.t. } j \leq |y| \leq j+1\}$$

It is then easy to see that $(V_\alpha \cup W_\beta)_{\alpha,\beta}$ is a uniformly locally finite covering of $\mathbf{R}^{n_1} \times \mathbf{R}^{n_2} \backslash B_0^1(R) \times B_0^2(R)$. As a consequence, there exists $K > 0$ such that for any nonnegative function $f \in \mathcal{D}_s(\mathbf{R}^n)$,

$$K \sum_{\alpha,\beta} \left(\int_{V_\alpha} f(x,y)dxdy + \int_{W_\beta} f(x,y)dxdy \right)$$
$$\leq \int_{X(R)} f(x,y)dxdy \leq \sum_{\alpha,\beta} \left(\int_{V_\alpha} f(x,y)dxdy + \int_{W_\beta} f(x,y)dxdy \right)$$

where $X(R) = \mathbf{R}^{n_1} \times \mathbf{R}^{n_2} \backslash B_0^1(R) \times B_0^2(R)$. Independently, for any $f \in \mathcal{D}_s(\mathbf{R}^n)$, $f \geq 0$,

$$\omega_{n_2-1} j^{n_2-1} \int_{B_i \times [j,j+1]} f(x,t)dxdt$$
$$\leq \int_{V_\alpha} f(x,t)dxdt \leq \omega_{n_2-1}(j+1)^{n_2-1} \int_{B_i \times [j,j+1]} f(x,t)dxdt$$

and

$$\omega_{n_1-1} i^{n_1-1} \int_{[i,i+1] \times B_j} f(t,y)dtdy$$
$$\leq \int_{W_\beta} f(t,y)dtdy \leq \omega_{n_1-1}(i+1)^{n_1-1} \int_{[i,i+1] \times B_j} f(t,y)dtdy$$

where $B_i = B^1_{a_i}(1)$ and $B_j = B^2_{b_j}(1)$. From now on, let $1 \leq q < n$ be given and let $q < p < nq/(n-q)$ be given. By [Ad, theorem 5.4] there exists $C > 0$ such that

(i) $\forall f \in C^\infty(\overline{B} \times [0,1])$,

$$\int_{B \times [0,1]} |f(x,t)|^p\, dx dt \leq C \left(\int_{B \times [0,1]} \left(|\nabla f(x,t)|^q + |f(x,t)|^q \right) dx dt \right)^{p/q}$$

where $B = B^1_0(1) \subset \mathbf{R}^{n_1}$

(ii) $\forall f \in C^\infty([0,1] \times \overline{B})$,

$$\int_{[0,1] \times B} |f(t,y)|^p\, dt dy \leq C \left(\int_{[0,1] \times B} \left(|\nabla f(t,y)|^q + |f(t,y)|^q \right) dt dy \right)^{p/q}$$

where $B = B^2_0(1) \subset \mathbf{R}^{n_2}$.

One then easily gets that for any $a \in \mathbf{R}^{n_1}$, any $j \in \mathbf{N}$, and any smooth function $f \in C^\infty(\overline{B}_a \times [j, j+1])$,

$$\int_{B_a \times [j,j+1]} |f(x,t)|^p\, dx dt \leq C \left(\int_{B_a \times [j,j+1]} \left(|\nabla f(x,t)|^q + |f(x,t)|^q \right) dx dt \right)^{p/q}$$

where $B_a = B^1_a(1) \subset \mathbf{R}^{n_1}$. Similarly, for any $b \in \mathbf{R}^{n_2}$, any $i \in \mathbf{N}$, and any $f \in C^\infty([i,i+1] \times \overline{B}_b)$,

$$\int_{[i,i+1] \times B_b} |f(t,y)|^p\, dt dy \leq C \left(\int_{[i,i+1] \times B_b} \left(|\nabla f(t,y)|^q + |f(t,y)|^q \right) dt dy \right)^{p/q}$$

where $B_b = B^2_b(1) \subset \mathbf{R}^{n_2}$. Let $u \in \mathcal{D}_s(\mathbf{R}^n)$. According to what we have said above, and noting that $p/q > 1$, one then has that

$$\int_{X(R)} |u(x,y)|^p\, dx dy$$

$$\leq \sum_{\alpha,\beta} \left(\int_{V_\alpha} |u(x,y)|^p\, dx dy + \int_{W_\beta} |u(x,y)|^p\, dx dy \right)$$

$$\leq \sum_{\alpha,\beta} \left(\omega_{n_2-1}\, (j+1)^{n_2-1} \int_{B_i \times [j,j+1]} |u(x,t)|^p\, dx dt \right.$$

$$\left. + \omega_{n_1-1}\, (i+1)^{n_1-1} \int_{[i,i+1] \times B_j} |u(t,y)|^p\, dt dy \right)$$

$$\leq C \sum_{\alpha,\beta} \left(\omega_{n_2-1}\, (j+1)^{n_2-1} \left(\int_{B_i \times [j,j+1]} \left(|\nabla u(x,t)|^q + |u(x,t)|^q \right) dx dt \right)^{p/q} \right.$$

$$+ \omega_{n_1-1} (i+1)^{n_1-1} \left(\int_{[i,i+1] \times B_j} (|\nabla u(t,y)|^q + |u(t,y)|^q) \, dt \, dy \right)^{p/q}$$

$$\leq C \min \left(\omega_{n_1-1}, \omega_{n_2-1} \right)^{1-p/q}$$

$$\times \sum_{\alpha,\beta} \left(\frac{(j+1)^{n_2-1}}{j^{(n_2-1)p/q}} \left(\int_{V_\alpha} (|\nabla u(x,y)|^q + |u(x,y)|^q) \, dx \, dy \right)^{p/q} \right.$$

$$+ \frac{(i+1)^{n_1-1}}{i^{(n_1-1)p/q}} \left(\int_{W_\beta} (|\nabla u(x,y)|^q + |u(x,y)|^q) \, dx \, dy \right)^{p/q} \right)$$

$$\leq C \min \left(\omega_{n_1-1}, \omega_{n_2-1} \right)^{1-p/q} max \left((\frac{[R]+1}{[R]^{p/q}})^{n_1-1}, (\frac{[R]+1}{[R]^{p/q}})^{n_2-1} \right)$$

$$\times \left(\sum_{\alpha,\beta} \int_{V_\alpha} (|\nabla u(x,y)|^q + |u(x,y)|^q) \, dx \, dy \right.$$

$$+ \int_{W_\beta} (|\nabla u(x,y)|^q + |u(x,y)|^q) \, dx \, dy \right)^{p/q}$$

$$\leq \frac{C}{K^{p/q}} \min \left(\omega_{n_1-1}, \omega_{n_2-1} \right)^{1-p/q} max \left((\frac{[R]+1}{[R]^{p/q}})^{n_1-1}, (\frac{[R]+1}{[R]^{p/q}})^{n_2-1} \right)$$

$$\times \left(\int_{X(R)} (|\nabla u(x,y)|^q + |u(x,y)|^q) \, dx \, dy \right)^{p/q}$$

As a consequence, we get that there exists a positive constant A such that for any real number $R \geq 1$ and any $u \in \mathcal{D}_s(\mathbf{R}^n)$,

$$\left(\int_{X(R)} |u(x,y)|^p \, dx \, dy \right)^{1/p} \leq A \, max \left((\frac{[R]+1}{[R]^{p/q}})^{n_1-1}, (\frac{[R]+1}{[R]^{p/q}})^{n_2-1} \right)^{1/p} \|u\|_{H_1^q}$$

By density such an inequality is valid for any $u \in H_{1,s}^q(\mathbf{R}^n)$. Independently, since $p/q > 1$, one has that

$$\lim_{R \to +\infty} max \left((\frac{[R]+1}{[R]^{p/q}})^{n_1-1}, (\frac{[R]+1}{[R]^{p/q}})^{n_2-1} \right)^{1/p} = 0$$

By [Ad, theorem 2.22] this ends the proof of the theorem.

References

[**Ad**] Adams, R.A., *Sobolev spaces*, Academic Press, 1978.

[**An1**] Anderson, M.T., *Ricci curvature bounds and Einstein metrics on compact manifolds*, Journal of the American Mathematical Society, 2, 1989, p. 455-490.

[**An2**] Anderson, M.T., *Convergence and rigidity of manifolds under Ricci curvature bounds*, Inventiones Mathematicae, 102, 1990, p. 429-445.

[**AC**] Anderson, M.T. and Cheeger, J., C^α *compactness for manifolds with Ricci curvature and injectivity radius bounded below*, Journal of Differential Geometry, 35, 1992, p. 265-281.

[**Au1**] Aubin, T., *Problèmes isopérimétriques et espaces de Sobolev*, Comptes Rendus de l'Académie des Sciences Paris, 280, 1974, p. 347-371.

[**Au2**] Aubin, T., *Espaces de Sobolev sur les variétés Riemanniennes*, Bulletin des Sciences Mathématiques, 100, 1976, p. 149-173.

[**Au3**] Aubin, T., *Problèmes isopérimétriques et espaces de Sobolev*, Journal of Differential Geometry, 11, 1976, p. 573-598.

[**Au4**] Aubin, T., *Equations différentielles non linéaires et problème de Yamabe concernant la courbure scalaire*, Journal de Mathématiques Pures et Appliquées, 55, 1976, p. 269-296.

[**Au5**] Aubin, T., *Meilleures constantes dans le théorème d'inclusion de Sobolev et un théorème de Fredholm non linéaire pour la transformation conforme de la courbure scalaire*, Journal of Functional Analysis, 32, 1979, p. 148-174.

[**Au6**] Aubin, T., *Nonlinear analysis on manifolds. Monge-Ampère equations*, Grundlehern der mathematischen Wissenschaften, Springer-Verlag, 252, 1982.

[**AE**] Aviles, P. and Escobar, J.F., *On the Sobolev quotient of an Einstein manifold*, Indiana University Mathematics Journal, 41, 1992, p. 435-438.

[**Ba**] Bakry, D., *L'hypercontractivité et son utilisation en théorie des semigroupes*, in Lectures on Probability theory, Lecture Notes in Mathematics, 1994, number 1581.

[**BaCLS**] Bakry, D., Coulhon, T., Ledoux, M. and Saloff-Coste, L., *Sobolev inequalities in disguise*, Preprint.

[**BaL**] Bakry, D. and Ledoux, M., *Sobolev inequalities and Myer's diameter theorem for an abstract markov generator*, To appear in Duke Mathelatical Journal.

[**Be**] Beckner, W., *Sharp Sobolev inequalities on the sphere and the Moser-Trudinger inequality*, Annals of Mathematics, 138, 1993, p. 213-242.

[**BeBG**] Berard, P., Besson, G. and Gallot, S., *Sur une inégalité isopérimétrique qui généralise celle de Paul Lévy-Gromov*, Inventiones Mathematicae, 80, 1985, p. 295-308.

[BeM] Berard, P. and Meyer, D., *Inégalités isopérimétriques et applications*, Annales Scientifiques de l'Ecole Normale Supérieure, 15, 1982, p.513-542.

[BeL] Berestycki, H. and Lions, P.L., *Existence of a ground state in nonlinear equations of the type Klein-Gordon, in Variational inequalities and complementarity theory and applications*, Wiley, N.Y., 1979.

[Bes] Besse, A.L., *Einstein manifolds*, Ergebnisse, Springer-Verlag, 10, 1987.

[BiV] Bidaut-Veron, M.F. and Veron, L., *Nonlinear elliptic equations on compact Riemannian manifolds and asymptotics of Emden equations*, Inventiones Mathematicae, 106, 1991, p. 489-539.

[Bl] Bliss, G.A., *An integral inequality*, J. London Math. Soc. 5, 1930, p. 40-46.

[Bo] Bourguignon, J.P., *The magic of Weitzenböck formulas*, in Variational methods, Progress in Nonlinear Differential Equations and Their Applications, Birkhäuser, 1990.

[Br] Bredon, G.E., *Introduction to compact transformation groups*, Academic Press, New York-London, 1972.

[BrL] Brézis, H. and Lieb E.H., *Sobolev inequalities with remainder terms*, Journal of Functional Analysis, 62, 1985, p. 73-86.

[Bu] Buser, P., *A note on the isoperimetric constant*, Annales Scientifiques de l'Ecole Normale Supérieure, 15, 1982, p. 213-230.

[CaGS] Caffarelli, L.A., Gidas, B. and Spruck, J., *Asymptotic symmetry and local behavior of semilinear elliptic equations with Sobolev growth*, Communications on Pure and Applied Mathematics, 17, 1989, p. 271-297.

[Can] Cantor, M., *Sobolev inequalities for Riemannian bundles*, Bulletin of the American Mathematical Society, 80, 1974, p. 239-243.

[Car1] Carron, G., *Inégalités isopérimétriques de Faber-Krahn et conséquences*, Prépublications de l'Institut Fourier, 1992, n220.

[Car2] Carron, G., *Inégalités isopérimétriques sur les variétés Riemanniennes*, Thèse de Doctorat de l'Université Joseph Fourier, 1994.

[ChYa] Chang S.Y.A. and Yang, P.C., *A perturbation result in prescribing scalar curvature on S^n*, Duke Mathematical Journal, 64, 1991, p. 27-69.

[Ch] Chavel, I., *Riemannian geometry: a modern introduction*, Cambridge tracts in Mathematics, Cambridge University press, 1993.

[ChGT] Cheeger, J , Gromov, M. and Taylor, M., *Finite propagation speed, kernel estimates for functions of the Laplace operator and the geometry of complete Riemannian manifolds*, Journal of Differential Geometry, 17, 1982, p. 15-53.

[ChL] Cheng, S.Y. and Li, P., *Heat kernel estimates and lower bound of eigenvalues*, Comment. Math. Helvetici, 56, 1981, p. 327-338.

[ChY] Cheng, S.Y. and Yau, S.T., *Differential equations on Riemannian man-*

ifolds and their geometric applications, Communications on Pure and Applied Mathematics, 28, 1975, p. 333-354.

[**Che**] Cherrier, P., *Une inégalité de Sobolev sur les variétés Riemanniennes*, Bulletin des Sciences Mathématiques, 103, 1979, p. 353-374.

[**CoGM**] Coleman, S., Glazer, V. and Martin, A., *Action minima among solutions to a class of euclidean scalar field equations*, Communications in Mathematical Physics, 58, 1978, p. 211-221.

[**CoI**] Cotsiolis A. and Iliopoulos, D., *Equations elliptiques non linéaires à croissance de Sobolev sur-critique*, Bulletin des Sciences Mathématiques, 119, 1995, p.419-431.

[**CoL**] Coulhon, T. and Ledoux, M., *Isopérimétrie, décroissance du noyau de la chaleur et transformations de Riesz: un contre-exemple*, Arkiv Mat., 32, 1994, p. 63-77.

[**CoS**] Coulhon, T. and Saloff-Coste, L., *Isopérimétrie pour les groupes et les variétés*, Revista Matematica Iberoamericana, 9, 1993, p. 293-314.

[**Cr1**] Croke, C.B., *Some isoperimetric inequalities and eigenvalue estimates*, Annales Scientifiques de l'Ecole Normale Supérieure, 13, 1980, p. 419-435.

[**Cr2**] Croke, C.B., *A sharp four dimensional isoperimetric inequality*, Comment. Math. Helvetici, 59, 1984, p. 187-192.

[**DK**] DeTurck, D.M. and Kazdan, J., *Some regularity theorems in Riemannian geometry*, Annales Scientifiques de l'Ecole Normale Supérieure, 14, 1981, p. 249-260.

[**Di**] Dieudonné, J., *Eléments d'analyse*, Gauthier-Villars, 1974.

[**Din**] Ding, W., *On a conformally invariant elliptic equation on R^n*, Communications in Mathematical Physics, 107, 1986, p. 331-335.

[**Es1**] Escobar, J.F., *Sharp constant in a Sobolev trace inequality*, Indiana University Mathematics Journal, 37, 1988, p. 687-698.

[**Es2**] Escobar, J.F., *Uniqueness theorems on conformal deformations of metrics, Sobolev inequalities and an eigenvalue estimate*, Communications on Pure and Applied Mathematics, 43, 1990, p. 857-883.

[**Fe**] Federer, H., *Geometric mesure theory*, New York, Springer-Verlag, 1969.

[**FeF**] Federer, H. and Fleming, W.H., *Normal integral currents*, Annals of Mathematics, 72, 1960, p. 458-520.

[**Fer**] Fernandez, J.L., *On the existence of Green's function in Riemannian manifolds*, Proceedings of the American Mathematical Society, 96, 1986, p. 284-286.

[**FlR**] Fleming, W.H. and Rishel, R., *An integral formula for total gradient variation*, Arch. Math., 11, 1960, p. 218-222.

[**Fo**] Fontenas, E., *Sur les constantes de Sobolev des variétés Riemanniennes* *compactes et les fonctions extrémales des sphères*, To appear in Bulletin des Sciences Mathématiques.

[Ga] Gagliardo,E., *Proprieta di alcune classi di funzioni in piu variabili*, Ric. Math., 7, 1958, p. 102-137.

[Gal1] Gallot, S., *Inégalités isopérimétriques, courbure de Ricci et invariants géométriques I*, Comptes Rendus de l'Académie des Sciences Paris, 296, 1983, p. 333-336.

[Gal2] Gallot, S., *Inégalités isopérimétriques, courbure de Ricci et invariants géométriques II*, Comptes Rendus de l'Académie des Sciences Paris, 296, 1983, p. 365-368.

[Gal3] Gallot, S., *Inégalités isopérimétriques et analytiques sur les variétés Riemanniennes*, Société Mathématique de France, Astérisque, 163-164, 1988, p. 31-91.

[Gal4] Gallot, S., *Isoperimetric inequalities based on integral norms of Ricci curvature*, Société Mathématique de France, Astérisque, 157-158, 1988, p. 191-216.

[GaHL] Gallot, S., Hulin, D. and Lafontaine, J., *Riemannian geometry*, 2nd edition, Universitext, Springer-Verlag, 1993.

[GiS] Gidas, B. and Spruck, J., *Global and local behavior of positive solutions of nonlinear elliptic equations*, Communications on Pure and Applied Mathematics, XXXIV, 1981, p. 525-598.

[GiNN] Gidas, B., Ni, W.M. and Nirenberg, L., *Symmetry and related properties via the maximum principle*, Communications in Mathematical Physics, 68, 1979, p. 209-243.

[GMGT] Glaser, V., Martin, A., Grosse, H. and Thirring, W., *A family of optimal conditions for the absence of bound states in a potential*, in Studies in Mathematical Physics, Princeton University Press, 1976, p. 169-194.

[Gr1] Grigor'yan, A.A., *On the existence of Green's function on a manifold*, Russian Math. Surveys, 38, 1983, p. 161-162.

[Gr2] Grigor'yan, A.A., *On the existence of positive fundamental solutions of the laplace equation on Riemannian manifolds*, Math. USSR Sbornik, 56, 1987, p. 349-358.

[Gro] Gromov, M., *Isoperimetric inequalities in Riemannian manifolds*, In Asymptotics Theory of Finite Dimensional Normed spaces, Lecture Notes Math. 1200, Springer-Verlag, 1986.

[GrLP] Gromov, M., Lafontaine, J. and Pansu, P., *Structures métriques pour les variétés Riemanniennes*, Cedic/fernand Nathan, 1981.

[H1] Hebey, E., *Changements de métriques conformes sur la sphère. Le problème de Nirenberg*, Bulletin des Sciences Mathématiques, 114, 1990, p. 215-242.

[H2] Hebey, E., *Courbure scalaire et géométrie conforme*, Journal of Geometry and Physics, 10, 1993, p. 345-380.

[H3] Hebey, E., *Scalar curvature on S^n and first spherical harmonics*, Journal of Differential Geometry and its Applications, 5, 1995, p. 71-78.

[H4] Hebey, E., *Optimal Sobolev inequalities on complete Riemannian manifolds with Ricci curvature bounded below and positive injectivity radius*, American Journal of Mathematics, 118, 1996, p. 291-300.

[HH] Hebey, E. and Herzlich, M., *Convergence of Riemannian manifolds - A status report*, Publications Mathématiques de l'Ecole Polytechnique, 1995, number 1096.

[HV1] Hebey, E. and Vaugon, M., *Meilleures constantes dans le théorème d'inclusion de Sobolev et multiplicité pour les problèmes de Nirenberg et Yamabe*, Indiana University Mathematics Journal, 41, 1992, p. 377-407.

[HV2] Hebey, E. and Vaugon, M., *Meilleures constantes dans le théorème d'inclusion de Sobolev*, Annales Inst. Henri Poincaré, Analyse non-linéaire, vol. 13, 1996, p. 57-93.

[HV3] Hebey, E. and Vaugon, M., *The best constant problem in the Sobolev embedding theorem for complete Riemannian manifolds*, Duke Mathematical Journal, 79, 1995, p. 235-279.

[HV4] Hebey, E. and Vaugon, M., *Effective L^p pinching for the concircular curvature*, To appear in Journal of Geometric Analysis.

[HV5] Hebey, E. and Vaugon, M., *Sobolev spaces in the presence of symmetries*, To appear in Journal de Mathématiques Pures et Appliquées.

[HoS] Hoffman, D. and Spruck, J., *Sobolev and isoperimetric inequalities for Riemannian submanifolds*, Communications on Pure and Applied Mathematics, vol. XXVII, 1974, p. 715-727.

[I1] Ilias, S., *Sur une inégalité de Sobolev*, Comptes Rendus de l'Académie des Sciences Paris, 294, 1982, p. 731-734.

[I2] Ilias, S., *Constantes explicites pour les inégalités de Sobolev sur les variétés Riemanniennes compactes*, Annales de l'Institut Fourier, 33, 1983, p. 151-165.

[Je] Jerison, D., *The Poincaré inequality for vector fields satisfying Hörmander's condition*, Duke Mathematical Journal, 53, 1986, p. 503-523.

[JK] Jost, J. and Karcher, H., *Geometrische methoden zur gewinnung von a-priori-schranken für harmonische abbildungen*, Manuscripta Math., 40, 1982, p. 27-77.

[Jo] Jost, J., *Riemannian geometry and geometric analysis*, Universitext, Springer-Verlag, 1995.

[Kl] Kleiner, B., *An isoperimetric comparison theorem*, Inventiones Mathematicae, 108, 1992, p. 37-47.

[Ko] Kobayashi, S., *Transformation groups in differential geometry*, Ergebnisse, 70, Springer-Verlag, 1972.

[KoN] Kobayashi, S. and Nomizu, K., Foundations of differential geometry, New York, Interscience Publishers, vol. I: 1963, vol.II: 1969.

[KuP] Kulkarni, R.S. and Pinkall, U., eds., *Conformal geometry*, Bonn, Vieweg and Sohn, Publications of the Max-Planck-Institut für Mathematik, 1988.

[LeP] Lee, J.M. and Parker, T.H., *The Yamabe problem*, Bulletin of the American Mathematical Society, 17, 1987, p. 37-91.

[Lef] Lelong-Ferrand, J., *Construction de modules de continuité dans le cas limite de Sobolev et applications à la géométrie différentielle*, Archive for Rational Mechanics and Analysis, 52, 1973, p. 297-311.

[Li] Li, P., *On the Sobolev constant and the p-spectrum of a compact Riemannian manifold*, Annales Scientifiques de l'Ecole Normale Supérieure, 13, 1980, p. 451-469.

[LiY] Li, P. and Yau, S.T., *Estimates of eigenvalues of a compact Riemannian manifold*, in Geometry of the Laplace operator, Proceedings of Symposia in Pure Mathematics, 36, 1980.

[LiZ] Li, Y.Y. and Zhu, M., *Sharp Sobolev trace inequality on Riemannian manifolds with boundary*, Preprint.

[Lic] Lichnerowicz, A., *Géométrie des groupes de transformation*, Paris, Dunod, 1958.

[Lie] Lieb, E.H., *Sharp constants in the Hardy-Littlewood-Sobolev and related inequalities*, Annals of Mathematics, 118, 1983, 349-374.

[Lio] Lions, J.L., *Equations différentielles opérationnelles et problèmes aux limites*, Lecture Notes in Mathematics, 61, Springer-Verlag, Berlin, 1961.

[Lio1] Lions, P.L., *Symétrie et compacité dans les espaces de Sobolev*, Journal of Functional Analysis, 49, 1982, p. 315-334.

[Lio2] Lions, P.L., *The concentration-compactness principle in the calculus of variations, I, II*, Revista Mathematica Iberoamericana, 1, 1985, p. 145-201 and 45-121.

[LiPT] Lions, P.L., Pacella, F. and Tricarico, M., *Best constants in Sobolev inequalities for functions vanishing on some part of the boundary and related questions*, Indiana University Mathematics Journal, 37, 1988, p. 301-324.

[MaS] Maheux, P. and Saloff-Coste, L., *Analyse sur les boules d'un opérateur sous-elliptique*, Mathematische Annalen, 303, 1995, p. 713-740.

[Maz] Maz'ja, V.G., *Sobolev spaces*, Springer-Verlag, 1985.

[Mi] Miranda, M., *Disguaglianze di Sobolev sulle ipersuperfici minimali*, Rendiconti del Seminario Matematico, 38, 1967, p. 69-79.

[MyS] Myers, S.B. and Steenrod, N., *The group of isometries of a Riemannian manifold*, Annals of Mathematics, 40, 1939, p. 400-416.

[Ni] Nirenberg, L., *On elliptic partial differential equations*, Ann. Scuola Norm. Sup. Pisa Sci. Fis. Mat., 13, 1959, p. 116-162.

[Ob] Obata, M., *The conjectures on conformal transformations of Riemannian manifolds*, Journal of Differential Geometry, 6, 1971, p. 247-258.

[Ok] Okikiolu, G.O., *Aspects of the theory of bounded integral operators in L^p-spaces*, Academic Press, N.Y., 1971.

[Os] Osserman, R., *The isoperimetric inequality*, Bulletin of the American Mathematical Society, 84, 1978, p. 1182-1238.

[Pa] Pansu, P., *Cohomologie L^p des variétés à courbure négative, cas du degré 1*, Rend. Sem. Mat. Univers. Politecn. Torino, Fascicolo Speciale, 1989, p. 95-120.

[Po] Pohozaev, S., *Eigenfunctions of the equation $\Delta u + \lambda f(u) = 0$*, Soviet Math. Dokl. 6, 1965, p. 1408-1411.

[Rc] R.C.A.M. Van der Vost, *Best constant for the embedding of the space $H^2 \cap H_0^1(\Omega)$ into $L^{2N/(N-4)}(\Omega)$*, Differential and Integral Equations, 6, 1993, p. 259-276.

[Ro] Rosen, G., *Minimum value for c in the Sobolev inequality*, SIAM J. Appl. Math., 21, 1971, p. 30-32.

[Sa1] Saloff-Coste, L., *Uniformly elliptic operators on Riemannian manifolds*, Journal of Differential Geometry, 36, 1992, p. 417-450.

[Sa2] Saloff-Coste, L., *A note on Poincaré, Sobolev, and Harnack inequalities*, Duke Mathematical Journal, International Mathematics Research Notices, 2, 1992, p. 27-38.

[Sc1] Schoen, R., *Conformal deformation of a Riemannian metric to constant scalar curvature*, Journal of Differential Geometry, 20, 1984, p. 479-495.

[Sc2] Schoen, R., *Variational theory for the total scalar curvature functional for Riemannian metrics and related topics*, in Topics in Calculus of Variations, Lecture Notes in Mathematics, 1365, Springer-Verlag, Berlin, 1989.

[ScY] Schoen, R. and Yau, S.T., *Conformally flat manifolds, Kleinian groups and scalar curvature*, Inventiones Mathematicae, 92, 1988, p. 47-71.

[So] Sobolev, S.L., *Sur un théorème d'analyse fonctionnelle*, Math. Sb. (N.S), 46, 1938, p. 471-496.

[Sp] Spivak, M., *A comprehensive introduction to differential geometry*, Boston, Publish or Perish, 1970.

[St] Strauss, W.A., *Existence of Solitary waves in higher dimensions*, Communications in Mathematical Physics, 55, 1977, p. 149-162.

[Ta] Talenti, G., *Best constants in Sobolev inequality*, Ann. di Matem. Pura ed Appl., 110, 1976, p. 353-372.

[Tr] Trudinger, N., *Remarks concerning the conformal deformation of Riemannian structures on compact manifolds*, Ann. Scuola Norm. Sup. Pisa, 22, 1968, p. 265-274.

[Va1] Varopoulos, N., *Small time Gaussian estimates of heat diffusion kernels. Part I: the semigroup technique*, Bulletin des Sciences Mathématiques, 113, 1989, p. 253-277.

[Va2] Varopoulos, N., *Potential theory and diffusion on Riemannian manifolds*, Conference in Harmonic Analysis in Honor of Antoni Zygmund, Wadsworth, Belmont, Calif, 1983.

[Va3] Varopoulos, N., *Hardy-Littlewood theory for semigroups*, Journal of Functional Analysis, 63, 1985, p. 240-260.

[Va4] Varopoulos, N., *Une généralisation du théorème de Hardy-Littlewood-Sobolev pour les espaces de Dirichlet*, Comptes Rendus de l'Académie des Sciences Paris, 299, 1984, p. 651-654.

[VSC] Varopoulos, N., Saloff-Coste, L. and Coulhon, T., *Analysis and geometry on groups*, Cambridge tracts in Mathematics, Cambridge University Press, 1992.

[Y] Yamabe, H., *On a deformation of Riemannian structures on compact manifolds*, Osaka Math. J., 12, 1960, p. 21-37.

[Ya] Yau, S.T., *Isoperimetric constants and the first eigenvalue of a compact manifold*, Annales Scientifiques de l'Ecole Normale Supérieure, 8, 1975, p. 487-507.

Notation index

For the reader's convenience we list the notations most frequently used in the text, and explain their meaning. When available, the number following the comma indicates the page where the notation is introduced.

$\mathcal{A}_q(M)$	space of Sobolev constants, 58
$Area_g(\partial\Omega)$	volume of $\partial\Omega$ w.r.t the metric induced by g
$\mathcal{B}_q(M)$	space of Sobolev constants, 58
$B_x(r)$	Ball of center x and radius r in (M, d_g)
$B_x^e(r)$	Euclidean ball of center x and radius r
$Card$	cardinal number
$C^\infty(M)$	space of smooth functions
$C_G^\infty(M)$	space of smooth G-invariant functions, 91
$det(g_{ij})$	determinant of (g_{ij})
dim	dimension
$diam_{(M,g)}$	diameter of (M, g)
d_e	Euclidean distance
d_g	Riemannian distance, 1
$dv(g)$	Riemannian volume element, 1
dx	Lebesgue's volume element
$\mathcal{D}(\Omega)$	smooth functions with compact support in Ω
exp_x	exponential map at x, 2
g	Riemannian metric, 1
g_{ij}	components of g
g^{ij}	components of the inverse matrix of (g_{ij})
$H_k^p(M)$	Sobolev space, 10
$\overset{\circ}{H}_k^p(M)$	Sobolev space, closure of $\mathcal{D}(M)$ in $H_k^p(M)$, 12
$H_{1,G}^p(M)$	Sobolev space for G-invariant functions, 90
$inj_{(M,g)}$	injectivity radius of (M, g), 2
(I_q)	Sobolev inequality, 58
$Isom_{(M,g)}$	group of isometries of (M, g)
$K_{(M,g)}$	sectional curvature of (M, g)
$K(n, q)$	optimal value for $\alpha_q(M)$, 61, 69
$L^p(M)$	Lebesgue's space
$L_G^p(M)$	Lebesgue's space for G-invariant functions, 91

Subject index

The following index does not contain the terms which we presume are well-known to the reader, even if they appear as definitions in the text (e.g. "Riemannian manifold"). For the other terms we indicate the main occurence only (i.e. the page where they are defined).

Vol. 1540: H. Komatsu (Ed.), Functional Analysis and Related Topics, 1991. Proceedings. XXI, 413 pages. 1993.

Vol. 1541: D. A. Dawson, B. Maisonneuve, J. Spencer, Ecole d´ Eté de Probabilités de Saint-Flour XXI - 1991. Editor: P. L. Hennequin. VIII, 356 pages. 1993.

Vol. 1542: J.Fröhlich, Th.Kerler, Quantum Groups, Quantum Categories and Quantum Field Theory. VII, 431 pages. 1993.

Vol. 1543: A. L. Dontchev, T. Zolezzi, Well-Posed Optimization Problems. XII, 421 pages. 1993.

Vol. 1544: M.Schürmann, White Noise on Bialgebras. VII, 146 pages. 1993.

Vol. 1545: J. Morgan, K. O'Grady, Differential Topology of Complex Surfaces. VIII, 224 pages. 1993.

Vol. 1546: V. V. Kalashnikov, V. M. Zolotarev (Eds.), Stability Problems for Stochastic Models. Proceedings, 1991. VIII, 229 pages. 1993.

Vol. 1547: P. Harmand, D. Werner, W. Werner, M-ideals in Banach Spaces and Banach Algebras. VIII, 387 pages. 1993.

Vol. 1548: T. Urabe, Dynkin Graphs and Quadrilateral Singularities. VI, 233 pages. 1993.

Vol. 1549: G. Vainikko, Multidimensional Weakly Singular Integral Equations. XI, 159 pages. 1993.

Vol. 1550: A. A. Gonchar, E. B. Saff (Eds.), Methods of Approximation Theory in Complex Analysis and Mathematical Physics IV, 222 pages, 1993.

Vol. 1551: L. Arkeryd, P. L. Lions, P.A. Markowich, S.R. S. Varadhan. Nonequilibrium Problems in Many-Particle Systems. Montecatini, 1992. Editors: C. Cercignani, M. Pulvirenti. VII, 158 pages 1993.

Vol. 1552: J. Hilgert, K.-H. Neeb, Lie Semigroups and their Applications. XII, 315 pages. 1993.

Vol. 1553: J.-L- Colliot-Thélène, J. Kato, P. Vojta. Arithmetic Algebraic Geometry. Trento, 1991. Editor: E. Ballico. VII, 223 pages. 1993.

Vol. 1554: A. K. Lenstra, H. W. Lenstra, Jr. (Eds.), The Development of the Number Field Sieve. VIII, 131 pages. 1993.

Vol. 1555: O. Liess, Conical Refraction and Higher Microlocalization. X, 389 pages. 1993.

Vol. 1556: S. B. Kuksin, Nearly Integrable Infinite-Dimensional Hamiltonian Systems. XXVII, 101 pages. 1993.

Vol. 1557: J. Azéma, P. A. Meyer, M. Yor (Eds.), Séminaire de Probabilités XXVII. VI, 327 pages. 1993.

Vol. 1558: T. J. Bridges, J. E. Furter, Singularity Theory and Equivariant Symplectic Maps. VI, 226 pages. 1993.

Vol. 1559: V. G. Sprindžuk, Classical Diophantine Equations. XII, 228 pages. 1993.

Vol. 1560: T. Bartsch, Topological Methods for Variational Problems with Symmetries. X, 152 pages. 1993.

Vol. 1561: I. S. Molchanov, Limit Theorems for Unions of Random Closed Sets. X, 157 pages. 1993.

Vol. 1562: G. Harder, Eisensteinkohomologie und die Konstruktion gemischter Motive. XX, 184 pages. 1993.

Vol. 1563: E. Fabes, M. Fukushima, L. Gross, C. Kenig, M. Röckner, D. W. Stroock, Dirichlet Forms. Varenna, 1992. Editors: G. Dell'Antonio, U. Mosco. VII, 245 pages. 1993.

Vol. 1564: J. Jorgenson, S. Lang, Basic Analysis of Regularized Series and Products. IX, 122 pages. 1993.

Vol. 1565: L. Boutet de Monvel, C. De Concini, C. Procesi, P. Schapira, M. Vergne. D-modules, Representation Theory, and Quantum Groups. Venezia, 1992. Editors: G. Zampieri, A. D'Agnolo. VII, 217 pages. 1993.

Vol. 1566: B. Edixhoven, J.-H. Evertse (Eds.), Diophantine Approximation and Abelian Varieties. XIII, 127 pages. 1993.

Vol. 1567: R. L. Dobrushin, S. Kusuoka, Statistical Mechanics and Fractals. VII, 98 pages. 1993.

Vol. 1568: F. Weisz, Martingale Hardy Spaces and their Application in Fourier Analysis. VIII, 217 pages. 1994.

Vol. 1569: V. Totik, Weighted Approximation with Varying Weight. VI, 117 pages. 1994.

Vol. 1570: R. deLaubenfels, Existence Families, Functional Calculi and Evolution Equations. XV, 234 pages. 1994.

Vol. 1571: S. Yu. Pilyugin, The Space of Dynamical Systems with the C⁰-Topology. X, 188 pages. 1994.

Vol. 1572: L. Göttsche, Hilbert Schemes of Zero-Dimensional Subschemes of Smooth Varieties. IX, 196 pages. 1994.

Vol. 1573: V. P. Havin, N. K. Nikolski (Eds.), Linear and Complex Analysis – Problem Book 3 – Part I. XXII, 489 pages, 1994.

Vol. 1574: V. P. Havin, N. K. Nikolski (Eds.), Linear and Complex Analysis – Problem Book 3 – Part II. XXII, 507 pages. 1994.

Vol. 1575: M. Mitrea, Clifford Wavelets, Singular Integrals, and Hardy Spaces. XI, 116 pages. 1994.

Vol. 1576: K. Kitahara, Spaces of Approximating Functions with Haar-Like Conditions. X, 110 pages. 1994.

Vol. 1577: N. Obata, White Noise Calculus and Fock Space. X, 183 pages. 1994.

Vol. 1578: J. Bernstein, V. Lunts, Equivariant Sheaves and Functors. V, 139 pages. 1994.

Vol. 1579: N. Kazamaki, Continuous Exponential Martingales and BMO. VII, 91 pages. 1994.

Vol. 1580: M. Milman, Extrapolation and Optimal Decompositions with Applications to Analysis. XI, 161 pages. 1994.

Vol. 1581: D. Bakry, R. D. Gill, S. A. Molchanov, Lectures on Probability Theory. Editor: P. Bernard. VIII, 420 pages. 1994.

Vol. 1582: W. Balser, From Divergent Power Series to Analytic Functions. X, 108 pages. 1994.

Vol. 1583: J. Azéma, P. A. Meyer, M. Yor (Eds.), Séminaire de Probabilités XXVIII. VI, 334 pages. 1994.

Vol. 1584: M. Brokate, N. Kenmochi, I. Müller, J. F. Rodriguez, C. Verdi, Phase Transitions and Hysteresis. Montecatini Terme, 1993. Editor: A. Visintin. VII. 291 pages. 1994.

Vol. 1585: G. Frey (Ed.), On Artin's Conjecture for Odd 2-dimensional Representations. VIII, 148 pages. 1994.

Vol. 1586: R. Nillsen, Difference Spaces and Invariant Linear Forms. XII, 186 pages. 1994.

Vol. 1587: N. Xi, Representations of Affine Hecke Algebras. VIII, 137 pages. 1994.

Vol. 1588: C. Scheiderer, Real and Étale Cohomology. XXIV, 273 pages. 1994.

Vol. 1589: J. Bellissard, M. Degli Esposti, G. Forni, S. Graffi, S. Isola, J. N. Mather, Transition to Chaos in Classical and Quantum Mechanics. Montecatini Terme, 1991. Editor: S. Graffi. VII, 192 pages. 1994.

Vol. 1590: P. M. Soardi, Potential Theory on Infinite Networks. VIII, 187 pages. 1994.

Vol. 1591: M. Abate, G. Patrizio, Finsler Metrics – A Global Approach. IX, 180 pages. 1994.

Vol. 1592: K. W. Breitung, Asymptotic Approximations for Probability Integrals. IX, 146 pages. 1994.

Vol. 1593: J. Jorgenson & S. Lang, D. Goldfeld, Explicit Formulas for Regularized Products and Series. VIII, 154 pages. 1994.

Vol. 1594: M. Green, J. Murre, C. Voisin, Algebraic Cycles and Hodge Theory. Torino, 1993. Editors: A. Albano, F. Bardelli. VII, 275 pages. 1994.

Vol. 1595: R.D.M. Accola, Topics in the Theory of Riemann Surfaces. IX, 105 pages. 1994.

Vol. 1596: L. Heindorf, L. B. Shapiro, Nearly Projective Boolean Algebras. X, 202 pages. 1994.

Vol. 1597: B. Herzog, Kodaira-Spencer Maps in Local Algebra. XVII, 176 pages. 1994.

Vol. 1598: J. Berndt, F. Tricerri, L. Vanhecke, Generalized Heisenberg Groups and Damek-Ricci Harmonic Spaces. VIII, 125 pages. 1995.

Vol. 1599: K. Johannson, Topology and Combinatorics of 3-Manifolds. XVIII, 446 pages. 1995.

Vol. 1600: W. Narkiewicz, Polynomial Mappings. VII, 130 pages. 1995.

Vol. 1601: A. Pott, Finite Geometry and Character Theory. VII, 181 pages. 1995.

Vol. 1602: J. Winkelmann, The Classification of Three-dimensional Homogeneous Complex Manifolds. XI, 230 pages. 1995.

Vol. 1603: V. Ene, Real Functions – Current Topics. XIII, 310 pages. 1995.

Vol. 1604: A. Huber, Mixed Motives and their Realization in Derived Categories. XV, 207 pages. 1995.

Vol. 1605: L. B. Wahlbin, Superconvergence in Galerkin Finite Element Methods. XI, 166 pages. 1995.

Vol. 1606: P.-D. Liu, M. Qian, Smooth Ergodic Theory of Random Dynamical Systems. XI, 221 pages. 1995.

Vol. 1607: G. Schwarz, Hodge Decomposition – A Method for Solving Boundary Value Problems. VII, 155 pages. 1995.

Vol. 1608: P. Biane, R. Durrett, Lectures on Probability Theory. VII, 210 pages. 1995.

Vol. 1609: L. Arnold, C. Jones, K. Mischaikow, G. Raugel, Dynamical Systems. Montecatini Terme, 1994. Editor: R. Johnson. VIII, 329 pages. 1995.

Vol. 1610: A. S. Üstünel, An Introduction to Analysis on Wiener Space. X, 95 pages. 1995.

Vol. 1611: N. Knarr, Translation Planes. VI, 112 pages. 1995.

Vol. 1612: W. Kühnel, Tight Polyhedral Submanifolds and Tight Triangulations. VII, 122 pages. 1995.

Vol. 1613: J. Azéma, M. Emery, P. A. Meyer, M. Yor (Eds.), Séminaire de Probabilités XXIX. VI, 326 pages. 1995.

Vol. 1614: A. Koshelev, Regularity Problem for Quasilinear Elliptic and Parabolic Systems. XXI, 255 pages. 1995.

Vol. 1615: D. B. Massey, Lê Cycles and Hypersurface Singularities. XI, 131 pages. 1995.

Vol. 1616: I. Moerdijk, Classifying Spaces and Classifying Topoi. VII, 94 pages. 1995.

Vol. 1617: V. Yurinsky, Sums and Gaussian Vectors. XI, 305 pages. 1995.

Vol. 1618: G. Pisier, Similarity Problems and Completely Bounded Maps. VII, 156 pages. 1996.

Vol. 1619: E. Landvogt, A Compactification of the Bruhat-Tits Building. VII, 152 pages. 1996.

Vol. 1620: R. Donagi, B. Dubrovin, E. Frenkel, E. Previato, Integrable Systems and Quantum Groups. VIII, 488 pages. 1996.

Vol. 1621: H. Bass, M. V. Otero-Espinar, D. N. Rockmore, C. P. L. Tresser, Cyclic Renormalization and Auto-morphism Groups of Rooted Trees. XXI, 136 pages. 1996.

Vol. 1622: E. D. Farjoun, Cellular Spaces, Null Spaces and Homotopy Localization. XIV, 199 pages. 1996.

Vol. 1623: H.P. Yap, Total Colourings of Graphs. VIII, 131 pages. 1996.

Vol. 1624: V. Brînzănescu, Holomorphic Vector Bundles over Compact Complex Surfaces. X, 170 pages. 1996.

Vol.1625: S. Lang, Topics in Cohomology of Groups. VII, 226 pages. 1996.

Vol. 1626: J. Azéma, M. Emery, M. Yor (Eds.), Séminaire de Probabilités XXX. VIII, 382 pages. 1996.

Vol. 1627: C. Graham, Th. G. Kurtz, S. Méléard, Ph. E. Protter, M. Pulvirenti, D. Talay, Probabilistic Models for Nonlinear Partial Differential Equations. X, 301 pages. 1996.

Vol. 1628: P.-H. Zieschang, An Algebraic Approach to Association Schemes. XII, 189 pages. 1996.

Vol. 1629: J. D. Moore, Lectures on Seiberg-Witten Invariants. VII, 105 pages. 1996.

Vol. 1630: D. Neuenschwander, Probabilities on the Heisenberg Group: Limit Theorems and Brownian Motion. VIII, 139 pages. 1996.

Vol. 1631: K. Nishioka, Mahler Functions and Transcendence. VIII, 185 pages.1996.

Vol. 1632: A. Kushkuley, Z. Balanov, Geometric Methods in Degree Theory for Equivariant Maps. VII, 136 pages. 1996.

Vol.1633: H. Aikawa, M. Essén, Potential Theory – Selected Topics. IX, 200 pages.1996.

Vol. 1634: J. Xu, Flat Covers of Modules. IX, 161 pages. 1996.

Vol. 1635: E. Hebey, Sobolev Spaces on Riemannian Manifolds. X, 116 pages, 1996.

General Remarks

Lecture Notes are printed by photo-offset from the master-copy delivered in camera-ready form by the authors. For this purpose Springer-Verlag provides technical instructions for the preparation of manuscripts.

Careful preparation of manuscripts will help keep production time short and ensure a satisfactory appearance of the finished book. The actual production of a Lecture Notes volume normally takes approximately 8 weeks.

Authors receive 50 free copies of their book. No royalty is paid on Lecture Notes volumes.

Authors are entitled to purchase further copies of their book and other Springer mathematics books for their personal use, at a discount of 33,3 % directly from Springer-Verlag.

Commitment to publish is made by letter of intent rather than by signing a formal contract. Springer-Verlag secures the copyright for each volume.

Addresses:

Professor A. Dold
Mathematisches Institut
Universität Heidelberg
Im Neuenheimer Feld 288
D-69120 Heidelberg
Federal Republic of Germany

Professor F. Takens
Mathematisch Instituut
Rijksuniversiteit Groningen
Postbus 800
NL-9700 AV Groningen
The Netherlands

Springer-Verlag, Mathematics Editorial
Tiergartenstr. 17
D-69121 Heidelberg
Federal Republic of Germany
Tel.: *49 (6221) 487-410